EXTRA CLASS RADIO AMATEUR FCC TEST MANUAL

by

MARTIN SCHWARTZ

Published by
AMECO PUBLISHING CORP.
220 East Jericho Turnpike
Mineola, New York 11501

EXTRA CLASS
RADIO AMATEUR
FCC TEST MANUAL

Copyright 1989
by the
Ameco Publishing Corp.

ISBN No. 0-912146-24-9

Library of Congress Catalog No. 86-72480

Printed in the United States of America

PREFACE

This Extra Class FCC Test Manual is part of a series of books published by the Ameco Publishing Corp. for the purpose of preparing individuals for the Federal Communications Commission Amateur Radio Operator examinations.

The questions and multiple choice answers in this manual have been issued by the Volunteer Examiner Coordinator's Committee under the supervision of the FCC. The 40 questions and multiple choice answers on the actual examination will be drawn from these questions. The questions in this book are divided into subelements. A specific number of questions (shown under each subelement heading), will be taken from each subelement. The discussions that follow the multiple choice answers were written by the author to give the reader sufficient background material to understand the reasons for choosing the proper multiple choice answer.

In those instances where this author feels that the correct multiple choice answer is complete and adequate for the proper understanding of the subject matter, there is no discussion; only the correct answer is indicated. In most of the questions, the discussions explain the correct multiple choice answers and give additional useful material that helps with the understanding of the questions and answers.

There are a few questions and/or multiple choice answers that are somewhat ambiguous. Since this author is not responsible for the questions or the multiple choice answers, they cannot be changed. However, in these few instances, the designated answer is given, together with a thorough explanation of the subject matter involved.

Although this guide deals with Element 4B of the Amateur Radio Operator's examinations, the Ameco Publishing Corp. publishes guides covering all the other amateur elements, as well as code courses for learning the International Morse Code. If additional theory background information is required, it is suggested that the Amateur Radio Theory Course, Cat. #102-01, be consulted. (See backcover for details). GOOD LUCK!

Martin Schwartz

TABLE OF CONTENTS

4BA-1A.1 What exclusive frequency privileges in the 80-meter band are authorized to Amateur Extra control operators?
A. 3525-3775 kHz B. 3500-3525 kHz C. 3700-3750 kHz D. 3500-3550 kHz

The answer is B. When we speak of the "80 meter" band, we generally mean the entire band of frequencies from 3500 to 4000 kHz. Sometimes, the entire band is referred to as the "75/80 meter" band. Sometimes, the phone portion of the band, which extends from 3750 to 4000 kHz, is referred to as the 75 meter band. In this question, the FCC reference to the "80 meter band" refers only to the lower half of the band that extends from 3500 to 3750 kHz. The frequencies, 3500 to 3525 kHz, are reserved exclusively for Extra class control operators. See question 4BA-1A.2 and the chart on page J-1.

4BA-1A.2 What exclusive frequency privileges in the 75-meter band are authorized to Amateur Extra control operators?
A. 3750-3775 kHz B. 3800-3850 kHz C. 3775-3800 kHz D. 3800-3825 kHz

The answer is A. The term, "75 meter" band, refers to the upper portion of the band (3750-4000 kHz) on which phone may be used. The lowest 25 kHz are reserved exclusively for Extra class control operators. See question 4BA-1A.1 and the chart on page J-1.

4BA-1A.3 What exclusive frequency privileges in the 40-meter band are authorized to Amateur Extra control operators?
A. 7000-7025 kHz B. 7000-7050 kHz C. 7025-7050 kHz D. 7100-7150 kHz

The answer is A. See chart on page J-1.

4BA-1A.4 What exclusive frequency privileges in the 20-meter band are authorized to Amateur Extra control operators?
A. 14.100-14.175 MHz and 14.150-14.175 MHz
B. 14.000-14.125 MHz and 14.250-14.300 MHz
C. 14.025-14.050 MHz and 14.100-14.150 MHz
D. 14.000-14.025 MHz and 14.150-14.175 MHz

The answer is D. See chart on page J-1.

4BA-1A.5 What exclusive frequency privileges in the 15-meter band are authorized to Amateur Extra control operators?
A. 21.000-21.200 MHz and 21.250-21.270 MHz
B. 21.050-21.100 MHz and 21.150-21.175 MHz
C. 21.000-21.025 MHz and 21.200-21.225 MHz
D. 21.000-21.025 MHz and 21.250-21.275 MHz

The answer is C. See chart on page J-1.

4BA-1B.1 What is a spurious emission or radiation?
A. As defined by Section 97.73, any emission or radiation falling outside the amateur band being used
B. As defined by Section 97.73, any emission or radiation other than

the fundamental that exceeds 25 microwatts, regardless of frequency
C. As defined by Section 97.73, any emission or radiation other than the fundamental that exceeds 10 microwatts, regardless of frequency
D. As defined by Section 97.73, any emission or radiation falling outside the amateur band that exceeds 25 microwatts

The answer is A.

4BA-1B.2 How much must the mean power of any spurious emission or radiation from an amateur transmitter be attenuated when the carrier frequency is below 30 MHz and the mean transmitted power is equal to or greater than 5 watts?
A. At least 30 dB below the mean power of the fundamental, and less than 25 mW
B. At least 40 dB below the mean power of the fundamental, and less than 50 mW
C. At least 30 dB below the mean power of the fundamental, and less than 50 mW
D. At least 40 dB below the mean power of the fundamental, and less than 25 mW

The answer is B. This rule holds for all equipment except those units built before April 15, 1977, or first marketed before Jan. 1, 1978.

4BA-1B.3 How much must the mean power of any spurious emission or radiation from an amateur transmitter be attenuated when the carrier frequency is above 30 MHz but below 225 MHz and the mean transmitted power is greater than 25 watts?
A. At least 30 dB below mean power of the fundamental
B. At least 40 dB below mean power of the fundamental
C. At least 50 dB below mean power of the fundamental
D. At least 60 dB below mean power of the fundamental

The answer is D. This rule holds for all equipment except those units built before April 15, 1977, or first marketed before January 1, 1978.

If the mean power is 25 watts or less, the mean power of any spurious radiation shall be at least 40 dB below the mean power of the fundamental, without exceeding 25 microwatts.

4BA-1B.4 What can the FCC require the licensee to do if any spurious radiation from an amateur station causes harmful interference to the reception of another radio station?
A. Reduce the spurious emissions to 0 dB below the fundamental
B. Observe quiet hours and pay a fine
C. Forfeit the station license and pay a fine
D. Eliminate or reduce the interference

The answer is D. Steps must be taken to eliminate the interference, in accordance with good engineering practice.

4BA-1C.1 What are the points of communication for an amateur station?
A. Other amateur stations only

B. Other amateur stations and other stations authorized by the FCC to communicate with amateurs
C. Other amateur stations and stations in the Personal Radio Service
D. Other amateur stations and stations in the Aviation or Private Land Mobile Radio Services

The answer is B. In addition, amateur radio stations may transmit one-way signals to receiving apparatus while in beacon operation or radio control operation.

4BA-1C.2 With which stations may an amateur station communicate?
A. Amateur, RACES and FCC Monitoring stations
B. Amateur stations and any other station authorized by the FCC to communicate with amateur stations
C. Amateur stations only
D. Amateur stations and US Government stations

The answer is B. Amateur stations may also communicate with U.S. Government stations for civil defense purposes, in emergencies and, on a temporary basis, for test purposes.

4BA-1C.3 Under what circumstances, if any, may an amateur station communicate with a non-amateur station?
A. Only during emergencies and when the Commission has authorized the non-amateur station to communicate with amateur stations
B. Under no circumstances
C. Only when the state governor has authorized that station to communicate with amateurs
D. Only during Public Service events in connection with REACT groups

The answer is A. See answer 4BA-1C.2.

4BA-1D.1 What rules must US citizens comply with when operating an Amateur Radio station in international waters?
A. The FCC rules contained in Part 15
B. The FCC rules contained in Part 97
C. The IARU rules governing international operation
D. There are no rules governing Amateur Radio operation in international waters

The answer is B.

4BA-1E.1 An Amateur Radio station is installed on board a ship or aircraft in a compartment separate from the main radio installation. What other conditions must the amateur operator comply with?
A. The Amateur Radio operation must be approved by the master of the ship of the captain of the aircraft
B. There must be an approved antenna switch included, so the amateur can use the ship or aircraft antennas, transmitting only when the main radios are not in use
C. The amateur station must have a power supply that is completely independent of the ship or aircraft power
D. The Amateur Radio operator must have an FCC Marine or Aircraft endorsement on his or her Amateur license

The answer is A.

4BA-1E.2 What types of licenses or permits are required before an amateur operator may transmit from a vessel registered in the US?
A. No amateur license is required outside of international waters
B. Any Amateur Radio license or Reciprocal Operating Permit issued by the FCC
C. Only amateurs holding General class or higher licenses may transmit from a vessel registered in the US
D. Only an Amateur Extra class licensee may operate aboard a vessel registered in the US
The answer is B. A reciprocal permit can be issued to an alien holding a valid amateur license from his government, provided there is a bilateral agreement between the United States and that government on a reciprocal basis by United States amateur operators.

4BA-2A.1 What is an FCC Reciprocal Operating Permit?
A. An FCC authorization to a holder of an amateur license issued by certain foreign governments to operate an Amateur Radio station in the United States and its possessions
B. An FCC permit to allow a United States licensed amateur to operate his station in a foreign nation, except Canada
C. An FCC permit allowing a foreign licensed amateur to handle traffic between the United States and the amateur's own nation, subject to FCC rules on traffic handling and third-party messages
D. An FCC permit to a commercial telecommunications company allowing that company to pay amateurs to handle traffic during emergencies
The answer is A. In order to issue the permit, there must be a bilateral agreement between the United States and the government of the alien seeking the permit that allows for such operation on a reciprocal basis by United States amateur radio operators.

4BA-2B.1 Who is eligible for an FCC Reciprocal Operating Permit?
A. Anyone holding a valid Amateur Radio license issued by a foreign government
B. Anyone holding a valid Amateur Radio license issued by a foreign government with which the United States has a reciprocal operating agreement, providing that person is not a United States citizen
C. Anyone who holds a valid Amateur Radio license issued by a foreign government with which the United States has a reciprocal operating agreement
D. Anyone other than a United States citizen who holds a valid Amateur Radio or shortwave listener's license issued by a foreign government
The answer is B. See answer 4BA-2A.1

4BA-2B.2 Under what circumstances, if any, is a US citizen holding a foreign Amateur Radio license eligible to obtain an FCC Reciprocal

Operating Permit?

A. A US Citizen is not eligible to obtain a Reciprocal Operating Permit for use in the United States

B. Only if the applicant brings his or her equipment from the foreign country

C. Only if that person is unable to qualify for a United States amateur license

D. If the applicant does not hold an FCC license as of the date of application, but had held a US amateur license other than Novice class less than 10 years before the date of application

The answer is A. A United States licensed amateur, who wishes to operate in a foreign country, is eligible for a permit in the foreign country, provided that the foreign country has a bilateral agreement with the United States for such operation on a reciprocal basis.

4BA–2C.1 What are the operator frequency privileges authorized by an FCC Reciprocal Operating Permit?

A. Those authorized to a holder of the equivalent United States amateur license, unless the FCC specifies otherwise by endorsement on the permit

B. Those that the holder of the Reciprocal Operating Permit would have if he were in his own country

C. Only those frequencies permitted to United States amateurs that the holder of the Reciprocal Operating Permit would have in his own country, unless the FCC specifies otherwise

D. Only those frequencies approved by the International Amateur Radio Union, unless the FCC specifies otherwise

The answer is C.

4BA–2D.1 How does an alien operator identify an Amateur Radio station when operating under an FCC Reciprocal Operating Permit?

A. By using only his or her own call

B. By using his or her own call, followed by the city and state in the United States or possessions closest to his or her present location

C. By using his or her own call, followed by the letter(s) and number indicating the United States call-letter district of his or her location at the time of the contact, with the city and state nearest the location specified once during each contact

D. By using his or her own call sign, followed by the serial number of the Reciprocal Operating Permit and the call-letter district number of his or her present location

The answer is C. His own call is followed by a slash for telegraphy, or the words "fixed", "portable", or "mobile" for radiotelephony; then the United States amateur call sign prefix letter(s) and numbers appropriate to the location of the station. In the case of phone, the identification must be in English.

4BA–3A.1 What is RACES?

A. An Amateur Radio network for providing emergency communications during long-distance athletic contests

B. The Radio Amateur Civil Emergency Service

C. The Radio Amateur Corps for Engineering Services

D. An Amateur Radio network providing emergency communications for transoceanic boat or aircraft races

The answer is B. RACES is a radiocommunication service conducted by volunteer licensed amateur radio operators, for providing emergency radiocommunications to local, regional, or state civil defense organizations.

4BA-3B.1 What is the purpose of RACES?

A. To provide civil-defense communications during emergencies

B. To provide emergency communications for transoceanic boat or aircraft races

C. To provide routine and emergency communications for long-distance athletic events

D. To provide routine and emergency communications for large- scale international events, such as the Olympic games

The answer is A. The Radio Amateur Civil Emergency Service provides for amateur radio operation for civil defense communications purposes only, during periods of local, regional or national civil emergencies, including any emergency which may necessitate invoking of the President's War Emergency Powers under the provisions of Section 606 of the Communications Act of 1934, as amended.

4BA-3C.1 With what other organization must an Amateur Radio station be registered before RACES registration is permitted?

A. The Amateur Radio Emergency Service

B. The US Department of Defense

C. A civil defense organization

D. The Amateur Auxiliary to the FCC Field Operations Bureau

The answer is C. An Amateur Radio station may not be operated in RACES unless a civil defense organization certifies that the amateur station is registered in the civil defense organization.

4BA-3D.1 Who may be the control operator of a RACES station?

A. Anyone who holds a valid FCC amateur operator's license other than Novice

B. Only an Amateur Extra class licensee

C. Anyone who holds an FCC Amateur Radio license other than Novice and is certified by a civil defense organization

D. Anyone who holds an FCC Amateur Radio license and is certified by a civil defense organization

The answer is D. The control operator must be certified as enrolled in a civil defense organization by that organization.

4BA-3E.1 What additional operator privileges are granted to an Amateur Extra class operator registered with RACES?

A. None   B. Permission to operate CW on 5167.5 kHz

C. Permission to operate an unattended HF packet radio station

D. Permission to operate on the 237-MHz civil defense band

The answer is A. The operator privileges in RACES are the same as those held by the operator in the Amateur Radio Service.

4BA-3F.1 What frequencies are normally available for RACES operation?

A. Only those frequencies authorized by the ARRL Section Emergency Coordinator

B. Only those frequencies listed in Section 97.8

C. Only transmitting frequencies in the top 25 kHz of each Amateur band

D. All frequencies available to the Amateur Radio Service

The answer is D. All frequencies and emissions authorized to the Amateur Radio Service are also available to RACES on a shared basis. See question 4BA-3G.1.

4BA-3G.1 What type of emergency can cause a limitation on the frequencies available for RACES operation?

A. An emergency in which the President invokes the War Emergency Powers under the provisions of the Communications Act of 1934

B. RACES operations must be confined to a single frequency band if the emergency is contained within a single state

C. RACES operations must be conducted on a VHF band if the emergency is confined to an area 25 miles or less in radius

D. The Red Cross may limit available frequencies if the emergency involves no immediate danger of loss of life

The answer is A. In the event of such an emergency, RACES operation is limited to specific frequencies and bands (as specified in 97.185 of the FCC Rules and Regulations).

4BA-3H.1 Which amateur stations may be operated in RACES?

A. Only Extra Class Amateur Radio stations

B. Any licensed Amateur Radio station except a station licensed to a Novice

C. Any licensed Amateur Radio station certified by the responsible civil defense organization

D. Any licensed Amateur Radio station other than a station licensed to a Novice, providing the station is certified by the responsible civil defense organization

The answer is C.

4BA-3H.2 What are the points of communications for amateur stations operated in RACES and certified by the responsible civil defense organization as registered with that organization?

A. Any RACES, civil defense, or Disaster Communications Service station

B. Any RACES stations and any FCC licensed amateur stations except stations licensed to Novices

C. Any FCC licensed amateur station or a station in the Disaster Communications Service

D. Any FCC licensed amateur station except stations licensed to Novices

The answer is A. In addition, RACES stations may communicate with any other station in any other service regulated by the Federal

Communications Commission, whenever such station is authorized by the Commission to exchange communications with stations in the Radio Amateur Civil Emergency Service.

4BA-3I.1 What are permissible communications in <u>RACES?</u>
A. Any communications concerning local traffic nets
B. Any communications concerning the Amateur Radio Emergency Service
C. Any communications concerning national defense and security or immediate safety of people and property that are authorized by the area civil defense organization
D. Any communications concerning national defense or security or immediate safety of people or property but only when a state of emergency has been declared by the President, the governor, or other authorized official, and then only so long as the state of emergency endures

    The answer is C. Other types of permissible communications are communications concerning public safety, maintenance of law and order, human suffering and need, dissemination of public information essential to civil defense training drills and brief transmissions for testing and adjusting of equipment.

4BA-4A.1 What are the purposes of the Amateur Satellite Service?
A. It is a radionavigation service using stations on earth satellites for the same purposes as those of the Amateur Radio Service
B. It is a radiocommunication service using stations on earth satellites for weather information
C. It is a radiocommunication service using stations on earth satellites for the same purpose as those of the Amateur Radio Service
D. It is a radiolocation service using stations on earth satellites for Amateur Radio operators engaged in satellite radar experimentation

    The answer is C.

4BA-4B.1 What are some frequencies available for <u>space operation?</u>
A. 7.0-7.1, 14.00-14.25, 21.00-21.45, 24.890-24.990, 28.00-29.70, 144-146, 435-438 and 24,000-24,050 MHz
B. 7.0-7.3, 21.00-21.45, 28.00-29.70, 144-146, 432-438 and 24,000-24,050 MHz
C. All frequencies available to the Amateur Radio Service, providing license-class, power and emission-type restrictions are observed
D. Only frequencies available to Amateur Extra Class licensees

    The answer is A. Those operating between 435 and 438 MHz shall not cause harmful interference to other services.

4BA-4C-1.1 What is the term used to describe an earth-to-space Amateur Radio communication that controls the functions of an amateur satellite?

A. Space operation   B. Telecommand operation
C. Earth operation   D. Control operation

    The answer is B. Telecommand operation is used to initiate, modify or terminate functions of a station in space operation.

4BA-4C-2.1 Which amateur stations are eligible for <u>telecommand operation</u>?

A. Any Amateur Radio licensee except Novice

B. Amateur Extra class licensees only

C. Telecommand operation is not permitted in the amateur satellite service

D. Any Amateur Radio station designated by the space station licensee

The answer is D. Telecommand operation means earth-to-space amateur radiocommunication to initiate, modify, or terminate functions of a station in space operation.

Any amateur radio station, designated by the licensee of a station in space operation, is eligible to conduct telecommand operation with the station in space operation, subject to the privileges of the control operator's class of license.

4BA-4D-1.1 What term is used to describe space-to-earth transmissions that communicate the results of measurements made by a station in space operation?

A. Data transmission

B. Frame check sequence

C. Telemetry

D. Telecommand operation

The answer is C. These measurements include those relating to the functions of the station.

4BA-4E-1.1 What is the term used to describe Amateur Radio communication from a station that is beyond the major portion of the earth's atmosphere?

A. EME

B. Exospheric operation

C. Downlink

D. Space operation

The answer is D. Space operation refers to a station that is beyond, or is intended to go beyond, or has been beyond the major portion of the earth's atmosphere.

4BA-4E-2.1 Which amateur stations are eligible for <u>space operation</u>?

A. Any licensee except Novice

B. General, Advanced and Extra class licensees only

C. Advanced and Extra class licensees only

D. Amateur Extra class licensees only

The answer is D. Space operation is space-to-earth, and space-to-space, amateur radio communication from a station which is beyond, is intended to go beyond, or has been beyond the major portion of the earth's atmosphere.

Amateur radio stations, licensed to Amateur Extra class operators, are eligible for space operation. The station licensee may permit any amateur radio operator to be the control operator, subject to the privileges of the control operator's class of license.

4BA-4E-4.1 When must the licensee of a station scheduled for <u>space operation</u> give the FCC written pre-space notification?

A. 3 months to 72 hours prior to initiating space operation

B. 6 months to 3 months prior to initiating space operation

C. 12 months to 3 months prior to initiating space operation

D. 27 months to 3 months prior to initiating space operation

The answer is D. Three notifications are required. The first one must be no less than 27 months prior to initiating space operation. The second one must be no less than 15 months prior to initiating space operation, and the third one must be no less than 3 months prior to initiating space operation.

4BA-4E-4.2 When must the licensee of a station in space operation give the FCC written in-space notification?
A. No later than 24 hours following initiation of space operation
B. No later than 72 hours following initiation of space operation
C. No later than 7 days following initiation of space operation
D. No later than 30 days following initiation of space operation

The answer is C. The notification must update the information contained in the pre-space operation notification.

4BA-4E-4.3 When must the licensee of a station in space operation give the FCC written post-space notification?
A. No later than 48 hours after termination is complete, under normal circumstances
B. No later than 72 hours after termination is complete, under normal circumstances
C. No later than 7 days after termination is complete, under normal circumstances
D. No later than 3 months after termination is complete, under normal circumstances

The answer is D. If the termination is ordered by the Commission, notification is required no later than twenty-four hours after termination is complete.

4BA-4F-1.1 What term describes earth-to-space-to-earth Amateur Radio communication by means of radio signals automatically retransmitted by a station in space operation?
A. Earth operation   B. ESE
C. Repeater operation   D. Auxiliary operation

The answer is A. See question 4BA-4F-2.1.

4BA-4F-2.1 Which amateur stations are eligible for earth operation?
A. Any Amateur Radio station
B. Amateur Extra class licensees only
C. Any licensee except Novice
D. A special license issued by the FCC is required before any Amateur Radio station is placed in earth operation

The answer is A. By EARTH OPERATION, we mean earth-to-space-to-earth amateur radiocommunication by means of radio signals automatically retransmitted by stations in space operation.

4BA-5A.1 What is a Volunteer-Examiner Coordinator?
A. An organization that is authorized to administer FCC Amateur Radio license examinations to candidates for the Novice license
B. An organization that is authorized to administer FCC Amateur Radio examinations for any class of license other than Novice

C. An organization that has entered into an agreement with the FCC to coordinate the efforts of Volunteer Examiners in preparing and administering examinations for Amateur Radio operator licenses

D. An organization that has entered into an agreement with the FCC to coordinate efforts of Volunteer Examiners in preparing and administering examinations for Amateur Radio operator licenses other than Novice

The answer is C.

4BA-5B.1 What are the requirements to be a VEC?

A. Be engaged in the manufacture and/or sale of amateur equipment or in the coordination of amateur activities throughout at least one call-letter district; and agree to abide by FCC Rules concerning administration of Amateur Radio examinations

B. Be organized at least partially for the purpose of furthering Amateur Radio; be at least regional in scope; and agree to abide by FCC Rules concerning coordination of Amateur Radio examinations

C. Be organized at least partially for the purpose of furthering Amateur Radio; be, at the most, county-wide in scope; and agree to abide by FCC Rules concerning administration of Amateur Radio examinations

D. Be engaged in a business related to Amateur Radio; and agree to administer Amateur Radio examinations in accordance with FCC Rules throughout at least one call-letter district

The answer is B. Also, a VEC must not accept compensation for services as a VEC and must not discriminate against any candidate.

4BA-5C.1 What are the functions of a VEC?

A. Accredit Volunteer Examiners; collect candidates' application forms, answer sheets and test results and forward the applications to the FCC; maintain pools of questions for Amateur Radio examinations; and perform other clerical tasks in accordance with FCC Rules

B. Assemble, print and sell FCC-approved examination forms; accredit Volunteer Examiners; collect candidates' answer sheets and forward them to the FCC; screen applications for completeness and authenticity; and perform other clerical tasks in accordance with FCC Rules

C. Accredit Volunteer Examiners; certify that examiners' equipment is type-accepted by the FCC; assemble, print and distribute FCC-approved examination forms; and perform other clerical tasks in accordance with FCC Rules

D. Maintain pools of questions for Amateur Radio examinations; administer code and theory examinations; score and forward the test papers to the FCC so that the appropriate license may be issued to each successful candidate

The answer is A. In addition, a VEC will coordinate exam scheduling.

4BA-5C.2 Where are the questions listed that must be used in written examinations?

A. In the appropriate VEC question pool

B. In PR Bulletin 1035C

C. In PL 97-259

D. In the appropriate FCC Report and Order

The answer is A. In the August 4, 1986 Report and Order, the FCC abolished PR Bulletins 1035 A, B, C and D and transferred the question pools to the VEC's.

4BA-5C.3 How is an Element 3(A) examination prepared?

A. By Advanced or Extra Class Volunteer Examiners or Volunteer-Examiner Coordinators selecting questions from the appropriate VEC question pool

B. By Volunteer-Examiner Coordinators selecting questions from the appropriate FCC bulletin

C. By Extra Class Volunteer Examiners selecting questions from the appropriate FCC bulletin

D. By the FCC selecting questions from the appropriate VEC question pool

The answer is A. Element 3(A) is the Technician Class theory examination covering general level (NOT General Class) theory and regulations.

4BA-5C.4 How is an Element 3(B) examination prepared?

A. By Extra Class Volunteer Examiners or Volunteer-Examiner Coordinators selecting questions from the appropriate VEC question pool

B. By Volunteer-Examiner Coordinators selecting questions from the appropriate FCC bulletin

C. By Extra Class Volunteer Examiners selecting questions from the appropriate FCC bulletin

D. By the FCC selecting questions from the appropriate VEC question pool

The answer is A. Element 3(B) is the General Class theory examination covering general level theory and regulations plus questions concerning the additional privileges of General Class licensees.

4BA-5C.5 How is an Element 4(A) examination prepared?

A. By Extra Class Volunteer Examiners or Volunteer-Examiner Coordinators selecting questions from the appropriate VEC question pool

B. By Volunteer-Examiner Coordinators selecting questions from the appropriate FCC bulletin

C. By Extra Class Volunteer Examiners selecting questions from the appropriate FCC bulletin

D. By the FCC selecting questions from the appropriate VEC question pool

The answer is A. In the August 4, 1986 Report and Order, the FCC abolished the various PR Bulletins relating to the test questions and transferred the test question pools to the VEC's.

4BA-5C.6 How is an Element 4(B) examination prepared?

A. By Extra Class Volunteer Examiners or Volunteer-Examiner Coordinators selecting questions from the appropriate VEC question pool
B. By Volunteer-Examiner Coordinators selecting questions from the appropriate FCC bulletin
C. By Extra Class Volunteer Examiners selecting questions from the appropriate FCC bulletin
D. By the FCC selecting questions from the appropriate VEC question pool

The answer is A. In the August 4, 1986 Report and Order, the FCC abolished the various PR Bulletins relating to the test questions and transferred the test question pools to the VEC's.

4BA-5D.1 What organization coordinates the dates and times for scheduling Amateur Radio examinations?
A. The FCC   B. A VEC   C. The IARU   D. Local radio clubs

The answer is B. A VEC is a Volunteer Examiner Coordinator who has entered into a written agreement with the FCC.

4BA-5E.1 Under what circumstances, if any, may a VEC refuse to accredit a person as a VE on the basis of membership in an Amateur Radio organization?
A. Under no circumstances
B. Only when the prospective VE is an ARRL member
C. Only when the prospective VE is not a member of the local Amateur Radio club
D. Only when the club is at least regional in scope

The answer is A.

4BA-5E.2 Under what circumstances, if any, may a VEC refuse to accredit a person as a VE on the basis of lack of membership in an Amateur Radio organization?
A. Under no circumstances
B. Only when the prospective VE is not an ARRL member
C. Only when the club is at least regional in scope
D. Only when the prospective VE is a not a member of the local Amateur Radio club giving the examinations

The answer is A.

4BA-5F.1 Under what circumstance, if any, may an organization engaged in the manufacture of equipment used in connection with Amateur Radio transmissions be a VEC?
A. Under no circumstances
B. If the organization's amateur-related sales are very small
C. If the organization is manufacturing very specialized amateur equipment
D. Only upon FCC approval that preventive measures have been taken to preclude any possible conflict of interest

The answer is D.

4BA-5F.2 Under what circumstances, if any, may a person who is an employee of a company that is engaged in the distribution of

equipment used in connection with Amateur Radio transmissions be a VE?

A. Under no circumstances
B. Only if the employee's work is not directly related to that part of the company involved in the manufacture or distribution of amateur equipment
C. Only if the employee has no financial interest in the company
D. Only if the employee is an Extra Class licensee

The answer is B.

4BA-5F.3 Under what circumstances, if any, may a person who owns a significant interest in a company that is engaged in the preparation of publications used in preparation for obtaining an amateur operator license be a VE?

A. Under no circumstances
B. Only if the organization's amateur related sales are very small
C. Only if the organization is publishing very specialized material
D. Only if the person is an Extra Class licensee

The answer is A. However, a person who does not normally communicate with that part of an entity engaged in the preparation or distribution of any publication used in preparation for obtaining amateur operator licenses, is eligible to be a VE.

4BA-5F.4 Under what circumstances, if any, may an organization engaged in the distribution of publications used in preparation for obtaining an amateur operator license be a VEC?

A. Under no circumstances
B. Only if the organization's amateur publishing business is very small
C. Only if the organization is selling the publication at cost to examinees
D. Only upon FCC approval that preventive measures have been taken to preclude any possible conflict of interest

The answer is D.

4BA-5G.1 Who may reimburse VEs and VECs for out-of-pocket expenses incurred in preparing, processing or administering examinations?

A. Examinees   B. FCC
C. ARRL   D. FCC and Examiners

The answer is A. VE's and VEC's may be reimbursed for out-of-pocket expenses for amateur station operator licenses above the Novice Class.

4BA-5G.2 What action must a VEC take against a VE who accepts reimbursement and fails to provide the annual expense certification?

A. Suspend the VE's accreditation for 1 year
B. Disaccredit the VE
C. Suspend the VE's accreditation and report the information to the FCC
D. Suspend the VE's accreditation for 6 months

The answer is B. The VEC must notify the FCC on January 31 of each year of the VE's that it has disaccredited for this reason.

4BA-5G.3 What type of expense records must be maintained by a VE who accepts reimbursement?
A. All out-of-pocket expenses and reimbursements from the examinees
B. All out-of-pocket expenses only
C. Reimbursements from examiners only
D. FCC reimbursements only
The answer is A. The VE and VEC must maintain these records for each examination session. They must certify to the FCC that all expenses for which reimbursement was obtained, were necessarily and prudently incurred.

4BA-5G.4 For what period of time must a VE maintain records of out-of-pocket expenses and reimbursements for each examination session for which reimbursement is accepted?
A. 1 year   B. 2 years   C. 3 years   D. 4 years
The answer is C. The records must be made available to the FCC upon request.

4BA-5G.5 By what date each year must a VE forward to the VEC a certification concerning expenses for which reimbursement was accepted?
A. December 15 following the year for which the reimbursement was accepted
B. January 15 following the year for which the reimbursement was accepted
C. April 15 following the year for which the reimbursement was accepted
D. October 15 following the year for which the reimbursement was accepted
The answer is B. The January 15th deadline allows the VEC two weeks time to gather these certifications and its own certification, and to forward them to the FCC by January 31st.

4BA-5G.6 For what type of services may a VE be reimbursed for out-of-pocket expenses?
A. Preparing, processing or administering examinations above the Novice class
B. Preparing, processing or administering examinations including the Novice class
C. A VE cannot be reimbursed for out-of-pocket expenses
D. Only for preparation of examination elements
The answer is A.

4BA-6A.1 What is an accredited Volunteer Examiner?
A. A General class radio amateur who is accredited by a VEC to administer examinations to applicants for Amateur Radio licenses
B. An Amateur Radio operator who is accredited by a VEC to administer examinations to applicants for Amateur Radio licenses
C. An Amateur Radio operator who administers examinations to

applicants for Amateur Radio licenses for a fee
D. An FCC staff member who tests volunteers who want to administer Amateur Radio examinations
The answer is B.

4BA-6A.2 What is an <u>accredited VE</u>?
A. A General class radio amateur who is accredited by a VEC to administer examinations to applicants for Amateur Radio licenses
B. An Amateur Radio operator who is accredited by a VEC to administer examinations to applicants for Amateur Radio licenses
C. An Amateur Radio operator who administers examinations to applicants for Amateur Radio licenses for a fee
D. An FCC staff member who tests volunteers who want to administer Amateur Radio examinations
The answer is B.

4BA-6B.1 What are the requirements for a <u>Volunteer Examiner</u> administering an examination for a Technician class operator license?
A. The Volunteer Examiner must be a Novice class licensee accredited by a Volunteer-Examiner Coordinator
B. The Volunteer Examiner must be an Advanced or Extra class licensee accredited by a Volunteer-Examiner Coordinator
C. The Volunteer Examiner must be an Extra Class licensee accredited by a Volunteer-Examiner Coordinator
D. The Volunteer Examiner must be a General class licensee accredited by a Volunteer-Examiner Coordinator
The answer is B.

4BA-6B.2 What are the requirements for a <u>Volunteer Examiner</u> administering an examination for a General class operator license?
A. The examiner must hold an Advanced class license and be accredited by a VEC
B. The examiner must hold an Extra Class license and be accredited by a VEC
C. The examiner must hold a General class license and be accredited by a VEC
D. The examiner must hold an Extra class license to administer the written test element, but an Advanced class examiner may administer the CW test element
The answer is B.

4BA-6B.3 What are the requirements for a <u>Volunteer Examiner</u> administering an examination for an Advanced class operator license?
A. The examiner must hold an Advanced class license and be accredited by a VEC
B. The examiner must hold an Extra class license and be accredited by a VEC
C. The examiner must hold a General class license and be accredited by a VEC
D. The examiner must hold an Extra class license to administer the written test element, but an Advanced class examiner may administer

the CW test element
The answer is B.

4BA-6B.4 What are the requirements for a <u>Volunteer Examiner</u> administering an examination for an Amateur <u>Extra class operator</u> license?

A. The examiner must hold an Advanced class license and be accredited by a VEC
B. The examiner must hold an Extra Class license and be accredited by a VEC
C. The examiner must hold a General class license and be accredited by a VEC
D. The examiner must hold an Extra class license to administer the written test element, but an Advanced class examiner may administer the CW test element
The answer is B.

4BA-6B.5 When is <u>VE accreditation</u> necessary?

A. Always in order to administer a Technician or higher class license examination
B. Always in order to administer a Novice or higher class license examination
C. Sometimes in order to administer an Advanced or higher class license examination
D. VE accreditation is not necessary in order to administer a General or higher class license examination
The answer is A.

4BA-6C.1 What is <u>VE accreditation</u>?

A. The process by which all Advanced and Extra class licensees are automatically given permission to conduct Amateur Radio examinations
B. The process by which the FCC tests volunteers who wish to coordinate Amateur Radio license examinations
C. The process by which the prospective VE requests his or her requirements for accreditation
D. The process by which each VEC makes sure its VEs meet FCC requirements to serve as Volunteer Examiners
The answer is D.

4BA-6C.2 What are the requirements for <u>VE accreditation</u>?

A. Hold an Advanced class license or higher; be at least 18 years old; not have any conflict of interest; and never had his or her amateur license suspended or revoked
B. Hold an Advanced class license or higher; be at least 16 years old; and not have any conflict of interest
C. Hold an Extra Class license or higher; be at least 18 years old; and be a member of ARRL
D. There are no requirements for accreditation, other than holding a General or higher class license
The answer is A.

4BA-6C.3 The services of which persons seeking to be VEs will not be accepted by the FCC?
A. Persons with Advanced class licenses
B. Persons being between 18 and 21 years of age
C. Persons who have ever had their amateur licenses suspended or revoked
D. Persons who are employees of the Federal Government
The answer is C.

4BA-6D.1 Under what circumstances, if any, may a person be compensated for services as a VE?
A. When the VE spends more than 4 hours at the test session
B. When the VE loses a day's pay to administer the exam
C. When the VE spends many hours preparing for the test session
D. Under no circumstances
The answer is D. While the VE may not be compensated for "services", the VE may be compensated for out-of-pocket expenses.

4BA-6D.2 How much money, if any, may a person accept for services as a VE?
A. None
B. Up to a half day's pay if the VE spends more than 4 hours at the test session
C. Up to a full day's pay if the VE spends more than 4 hours preparing for the test session
D. Up to $50 if the VE spends more than 4 hours at the test session
The answer is A. See answer 4BA-6D.1.

4BA-7A-1.1 What is an Element 1(A) examination intended to prove?
A. The applicant's ability to send and receive Morse code at 5 WPM
B. The applicant's ability to send and receive Morse code at 13 WPM
C. The applicant's knowledge of Novice class theory and regulations
D. The applicant's ability to send and receive Morse code at 20 WPM
The answer is A.

4BA-7A-1.2 What is an Element 1(B) examination intended to prove?
A. The applicant's knowledge of Novice class theory and regulations
B. The applicant's knowledge of General class theory and regulations
C. The applicant's ability to send and receive Morse code at 5 WPM
D. The applicant's ability to send and receive Morse code at 13 WPM
The answer is D.

4BA-7A-1.3 What is an Element 1(C) examination intended to prove?
A. The applicant's ability to send and receive Morse code at 20 WPM
B. The applicant's knowledge of Amateur Extra class theory and regulations
C. The applicant's ability to send and receive Morse code at 13 WPM
D. The applicant's ability to send and receive Morse code at 5 WPM
The answer is A.

4BA-7A-1.4 What is Examination Element 2?

A. The 5-WPM amateur Morse code examination
B. The 13-WPM amateur Morse code examination
C. The written examination for the Novice class operator license
D. The written examination for the Technician class operator license
The answer is C. This is the theory test for the Novice Class.

4BA-7A-1.5 What is Examination Element 3(A)?
A. The 5-WPM amateur Morse code examination
B. The 13-WPM amateur Morse code examination
C. The written examination for the Technician class operator license
D. The written examination for the General class operator license
The answer is C. This element contains at least 25 questions.

4BA-7A-1.6 What is Examination Element 3(B)?
A. The 5-WPM amateur Morse code examination
B. The 13-WPM amateur Morse code examination
C. The written examination for the Technician class operator license
D. The written examination for the General class operator license
The answer is D. This element contains at least 25 questions.

4BA-7A-1.7 What is Examination Element 4(A)?
A. The written examination for the Technician class operator license
B. The 20-WPM amateur Morse code examination
C. The written examination for the Advanced class operator license
D. The written examination for the Amateur Extra class operator license
The answer is C.

4BA-7A-1.8 What is Examination Element 4(B)?
A. The written examination for the Technician class operator license
B. The 20-WPM amateur Morse code examination
C. The written examination for the Advanced class operator license
D. The written examination for the Amateur Extra class operator license
The answer is D.

4BA-7A-2.1 Who must prepare Examination Element 1(B)?
A. Extra class licensees serving as Volunteer Examiners, or Volunteer-Examiner Coordinators
B. Advanced class licensees serving as Volunteer Examiners, or Volunteer-Examiner Coordinators
C. The FCC
D. The Field Operations Bureau
The answer is A.

4BA-7A-2.2 Who must prepare Examination Element 1(C)?
A. The FCC   B. The Field Operations Bureau
C. Advanced class licensees serving as Volunteer Examiners, or Volunteer-Examiner Coordinators
D. Extra class licensees serving as Volunteer Examiners, or Volunteer-Examiner Coordinators
The answer is D.

4BA-7A-2.3 Who must prepare <u>Examination Element 3(A)</u>?
A. Advanced or Extra class licensees serving as Volunteer Examiners, or Volunteer-Examiner Coordinators
B. The FCC
C. The Field Operations Bureau
D. Advanced or General class licensees serving as Volunteer Examiners, or Volunteer-Examiner Coordinators
    The answer is A.

4BA-7A-2.4 Who must prepare <u>Examination Element 3(B)</u>?
A. Extra class licensees serving as Volunteer Examiners, or Volunteer-Examiner Coordinators
B. The FCC
C. The Field Operations Bureau
D. Advanced or General class licensees serving as Volunteer Examiners, or Volunteer-Examiner Coordinators
    The answer is A.

4BA-7A-2.5 Who must prepare <u>Examination Element 4(A)</u>?
A. Advanced or Extra class licensees serving as Volunteer Examiners, or Volunteer-Examiner Coordinators
B. The FCC
C. The Field Operations Bureau
D. Extra class licensees serving as Volunteer Examiners, or Volunteer-Examiner Coordinators
    The answer is D.

4BA-7A-2.6 Who must prepare <u>Examination Element 4(B)</u>?
A. Advanced or Extra class licensees serving as Volunteer Examiners, or Volunteer-Examiner Coordinators
B. The FCC
C. The Field Operations Bureau
D. Extra class licensees serving as Volunteer Examiners, or Volunteer-Examiner Coordinators
    The answer is D.

4BA-7B.1 What examination elements are required for an Amateur Extra class operator license?
A. 1(C) and 4(B)
B. 3(B), 4(A) and 4(B)
C. 1(B), 2, 3(A), 3(B), 4(A) and 4(B)
D. 1(C), 2, 3(A), 3(B), 4(A) and 4(B)
    The answer is D. See answers 4BA-7A-1.1 through 4BA-7A-1.8.

4BA-7B.2 What examination elements are required for an Advanced class operator license?
A. 1(A), 2, 3(A), 3(B) and 4(A)   B. 1(B), 3(A) and 3(B)
C. 1(B) and 4(A)   D. 1(B), 2, 3(A), 3(B) and 4(A)
    The answer is D. See answers 4BA-7A-1.1 through 4BA-7A-1.8.

4BA-7B.3 What examination elements are required for a General class operator license?

A. 1(B), 2, 3(A) and 3(B)   B. 1(A), 2, 3(A) and 3(B)
C. 1(A), 3(A) and 3(B)   D. 1(B), 3(A) and 3(B)

The answer is A. See answers 4BA-7A-1.1 through 4BA-7A-1.8.

4BA-7B.4 What examination elements are required for a Technician class operator license?

A. 1(A) and 2B   B. 1(A) and 3(A)
C. 1(A), 2 and 3(A)   D. 2 and 3(A)

The answer is C. See answers 4BA-7A-1.1 through 4BA-7A-1.8.

4BA-7C.1 What examination credit must be given to an applicant who holds a valid Novice class operator license?

A. Credit for successful completion of Elements 1(A) and 2
B. Credit for successful completion of Elements 1(B) and 3(A)
C. Credit for successful completion of Elements 1(B) and 2
D. Credit for successful completion of Elements 1(A) and 3(A)

The answer is A. An applicant who has already passed certain elements need not repeat them in going for a higher class of license.

4BA-7C.2 What examination credit must be given to an applicant who holds a valid Technician class operator license issued after March 20, 1987?

A. Credit for successful completion of Elements 1(A) and 2
B. Credit for successful completion of Elements 1(A), 2 and 3(A)
C. Credit for successful completion of Elements 1(B), 2 and 3(A)
D. Credit for successful completion of Elements 1(B), 3(A) and 3(B)

The answer is B. See answer 4BA-7C.1. Prior to March 20, 1987 the same 50 question Element 3 examination was given to both Technician and General Class applicants. As of March 20, 1987, Element 3 was broken up into a 25 question test, called Element 3(A), for the Technician Class applicants and a 25 question test, called Element 3(B), for the General Class applicants. See question 4BA-7C.3.

4BA-7C.3 What examination credit must be given to an applicant who holds a valid Technician class operator license issued before March 21, 1987?

A. Credit for successful completion of Elements 1(A), 2 and 3(B)
B. Credit for successful completion of Elements 1(A), 2, 3(A) and 3(B)
C. Credit for successful completion of Elements 1(B), 2, 3(A) and 4(A)
D. Credit for successful completion of Elements 1(B), 3(A) and 3(B)

The answer is B. See question 4BA-7C.2. If the applicant's Technician Class license was issued before March 20, 1987, he/she took the 50 question test that contained theory information for both the Technician and General Class licenses. He/she therefore receives credit for both the Technician and General Class theory elements.

4BA-7C.4 What examination credit must be given to an applicant who holds a valid General class operator license?

A. Credit for successful completion of Elements 1(B), 2, 3(A), 3(B) and 4(A)

B. Credit for successful completion of Elements 1(A), 3(A), 3(B) and 4(A)
C. Credit for successful completion of Elements 1(A), 2, 3(A), 3(B) and 4(B)
D. Credit for successful completion of Elements 1(B), 2, 3(A) and 3(B)
   The answer is D.

4BA-7C.5 What examination credit must be given to an applicant who holds a valid Advanced class operator license?
A. Credit for successful completion of Element 4(A)
B. Credit for successful completion of Elements 1(B) and 4(A)
C. Credit for successful completion of Elements 1(B), 2, 3(A), 3(B) and 4(A)
D. Credit for successful completion of Elements 1(C), 3(A), 3(B), 4(A) and 4(B)
   The answer is C.

4BA-7C.6 What examination credit, if any, may be given to an applicant who holds a valid amateur operator license issued by another country?
A. Credit for successful completion of any elements that may be identical to those required for U.S. licensees
B. No credit
C. Credit for successful completion of Elements 1(A), 1(B) and 1(C)
D. Credit for successful completion of Elements 2, 3(A), 3(B), 4(A) and 4(B)
   The answer is B.

4BA-7C.7 What examination credit, if any, may be given to an applicant who holds a valid amateur operator license issued by any other United States government agency than the FCC?
A. No credit
B. Credit for successful completion of Elements 1(A), 1(B) or 1(C)
C. Credit for successful completion of Elements 4(A) and 4(B)
D. Credit for successful completion of Element 1(C)
   The answer is A.

4BA-7C.8 What examination credit must be given to an applicant who holds a valid FCC commercial radiotelegraph license?
A. No credit
B. Credit for successful completion of element 1(B) only
C. Credit for successful completion of elements 1(A), 1(B) or 1(C)
D. Credit for successful completion of element 1(A) only
   The answer is C. A person who applies for an amateur operator license will be given credit for any telegraphy element if that person holds a Commercial Radiotelegraph Operator License or Permit issued by the Federal Communications Commission, or has held one within 5 years of the Commission's receipt of that persons application for an amateur operator license.

4BA-7C.9 What examination credit must be given to the holder of a valid Certificate of Successful Completion of Examination?
A. Credit for previously completed written examination elements only

B. Credit for the code speed associated with the previously completed telegraphy examination elements only
C. Credit for previously completed written and telegraphy examination elements only
D. Credit for previously completed commercial examination elements only

The answer is C. For purposes of examination credit, this certificate is valid for a period of one year from the date of issuance.

4BA-7D.1 Who determines where and when examinations for amateur operator licenses are to be administered?
A. The FCC
B. The Section Manager
C. The applicants
D. The administering Volunteer Examiner Team

The answer is D. Each examination for an amateur radio license must be administered at a location and time specified by the examiner(s).

4BA-7D.2 Where must the examiners be and what must they be doing during an examination?
A. The examiners must be present and observing the candidate(s) throughout the entire examination
B. The examiners must be absent to allow the candidate(s) to complete the entire examination in accordance with the traditional honor system
C. The examiners must be present to observe the candidate(s) throughout the administration of telegraphy examination elements only
D. The examiners must be present to observe the candidate(s) throughout the administration of written examination elements only

The answer is A.

4BA-7D.3 Who is responsible for the proper conduct and necessary supervision during an examination?
A. The VEC
B. The FCC
C. The administering Volunteer Examiners
D. The candidates and the administering Volunteer Examiners

The answer is C.

4BA-7D.4 What should an examiner do when a candidate fails to comply with the examiner's instructions?
A. Warn the candidate that continued failure to comply with the examiner's instructions will result in termination of the examination
B. Immediately terminate the examination
C. Allow the candidate to complete the examination, but refuse to issue a Certificate of Successful Completion of Examination for any elements passed by fraudulent means
D. Immediately terminate the examination and report the violation to federal law enforcement officials

The answer is B.

4BA-7D.5 What must the candidate do at the completion of the examination?
A. Complete a brief written evaluation of the examination session
B. Return all test papers to the examiners
C. Return all test papers to the examiners and wait for them to be graded before leaving the examination site
D. Pay the registration fee
The answer is B. Answer C has some validity since the test papers must be graded immediately upon the completion of an examination element. In addition, the applicant must be notified whether he/she passed or not. This is in accordance with Rule 97.28(c)(d)(e).

4BA-7E.1 When must the test papers be graded?
A. Within 5 days of completion of an examination element
B. Within 30 days of completion of an examination element
C. Immediately upon completion of an examination element
D. Within 10 days of completion of an examination element
The answer is C.

4BA-7E.2 Who must grade the test papers?
A. The ARRL
B. The administering Volunteer Examiners
C. The Volunteer-Examiner Coordinator
D. The FCC
The answer is B.

4BA-7E.3 How do the examiners inform a candidate who does not score a passing grade?
A. Give the percentage of the questions answered correctly and return the application to the candidate
B. Give the percentage of the questions answered incorrectly and return the application to the candidate
C. Tell the candidate that he or she failed and return the application to the candidate
D. Show how the incorrect answers should have been answered and give a copy of the corrected answer sheet to the candidate
The answer is A.

4BA-7E.4 What must the examiners do when the candidate scores a passing grade?
A. Give the percentage of the questions answered correctly and return the application to the candidate
B. Tell the candidate that he or she passed
C. Issue the candidate an operator license
D. Issue the candidate a Certificate of Successful Completion of Examination for the appropriate exam element(s)
The answer is D. This Certificate must bear the VEC-issued examination identification identifier code. Within one year, this Certificate may also be used for examination credit for Elements 1(A), 1(B) or 1(C).

4BA-7E.5 Within what time limit after administering an exam must the examiners submit the applications and test papers from successful candidates to the VEC?
A. Within 10 days   B. Within 15 days
C. Within 30 days   D. Within 90 days
   The answer is A.

4BA-7E.6 To whom do the examiners submit successful candidates' applications and test papers?
A. To the candidate   B. To the coordinating VEC
C. To the local radio club   D. To the regional Section Manager
   The answer is B.

4BA-7F.1 When an applicant passes an examination to upgrade his or her operator license, under what authority may he or she be the control operator of an amateur station with the privileges of the higher operator class?
A. That of the Certificate of Successful Completion of Examination issued by the VE Team that administered the examination
B. That of the ARRL
C. Applicants already licensed in the Amateur Radio Service may not use their newly earned privileges until they receive their permanent amateur station and operator licenses
D. Applicants may only use their newly earned privileges during emergencies pending issuance of their permanent amateur station and operator licenses
   The answer is A. He may do this for a period of one year, provided that the applicant retains the Certificate(s) for Successful Completion of the examination(s) at the station location, and provided that the applicant uses the identifier code of the new class of license for which the applicant has qualified (KT for Technician Class, AG for General Class, AA for Advanced Class and AE for Amateur Extra Class) as a suffix to the present call sign, and provided that the FCC has not yet acted upon the application for a higher class of license.

4BA-7F.2 What is a Certificate of Successful Completion of Examination?
A. A document printed by the FCC
B. A document required for already licensed applicants operating with privileges of an amateur operator class higher than that of their permanent amateur operator licenses
C. A document a candidate may use for an indefinite period of time to receive credit for successful completion of any written element
D. A permanent Amateur Radio station and operator license certificate issued to a newly-upgraded licensee by the FCC within 90 days of the completion of the examination
   The answer is B. It is a document indicating that the applicant has passed a particular element.

4BA-7F.3 How long may a successful applicant operate a station under Section 97.35 with the rights and privileges of the higher operator class for which the applicant has passed the appropriate

examinations?

A. 30 days or until issuance of a permanent operator and station license, whichever comes first

B. 3 months or until issuance of the permanent operator and station license, whichever comes first

C. 6 months or until issuance of the permanent operator and station license, whichever comes first

D. 1 year or until issuance of the permanent operator and station license, whichever comes first

The answer is D.

4BA-7F.4 How must the station call sign be amended when operating under the temporary authority authorized by Section 97.35?

A. The applicant must use an identifier code as a prefix to his or her present call sign, e.g., when using voice; "interim AE KA1MJP"

B. The applicant must use an identifier code as a suffix to his or her present call sign, e.g., when using voice; "KA1MJP temporary AE"

C. By adding after the call sign, when using voice, the phrase "operating temporary Technician, General, Advanced or Extra"

D. By adding to the call sign, when using CW, the slant bar followed by the letters T, G, A or E

The answer is B. See answer 4BA-7F.1.

SUBELEMENT 4BB
OPERATING PROCEDURES
(4 questions)

4BB-1A.1 What is an ascending pass for an amateur satellite?
A. A pass from west to east   B. A pass from east to west
C. A pass from south to north   D. A pass from north to south
The answer is C.

4BB-1A.2 What is a descending pass for an amateur satellite?
A. A pass from north to south   B. A pass from west to east
C. A pass from east to west   D. A pass from south to north
The answer is A.

4BB-1A.3 What is the period of an amateur satellite?
A. An orbital arc that extends from 60 degrees west longitude to 145 degrees west longitude
B. The point on an orbit where satellite height is minimum
C. The amount of time it takes for a satellite to complete one orbit
D. The time it takes a satellite to travel from perigee to apogee
The answer is C.

4BB-1B.1 What is Mode A in an amateur satellite?
A. Operation through a 10-meter receiver on a satellite that retransmits on 2 meters
B. The lowest frequency used in Phase 3 transponders
C. The highest frequency used in Phase 3 translators
D. Operation through a 2-meter receiver on a satellite that retransmits on 10 meters
The answer is D. The uplink 2 meter frequencies are in the range of 144.850 to 144.950 MHz and the 10 meter downlink frequencies are in the range of 29.4 to 29.5 MHz.

A typical satellite link is shown in Figure 4BB-1B.1. Station A transmits to the satellite on a frequency in a VHF band (such as 2 meters). At the satellite, the signal is converted and retransmitted in a different amateur band (such as 10 meters). The signal traveling to the satellite is called the "uplink", the return signal is called the "downlink". The station at B listens to the downlink frequency and

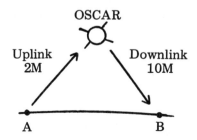

Fig. 4BB-1B.1. A typical satellite system; other bands are possible.

answers on the same VHF uplink. The result is a form of duplex operation. The two stations can thus remain in contact for as long as the satellite remains in position between them - as much as twenty minutes with some current satellites.

4BB-1B.2 What is <u>Mode B</u> in an amateur satellite?
A. Operation through a 10-meter receiver on a satellite that retransmits on 2 meters
B. Operation through a 70-centimeter receiver on a satellite that retransmits on 2 meters
C. The beacon output
D. A codestore device used to record messages

The answer is B. The 70 centimeter uplink frequencies are in the range of 432.125 to 432.175 MHz, and the 2 meter downlink frequencies are in the range of 145.925 to 145.975 MHz.

4BB-1B.3 What is <u>Mode J</u> in an amateur satellite?
A. Operation through a 70-centimeter receiver on a satellite that retransmits on 2 meters
B. Operation through a 2-meter receiver on a satellite that retransmits on 70 centimeters
C. Operation through a 2-meter receiver on a satellite that retransmits on 10 meters
D. Operation through a 70-centimeter receiver on a satellite that retransmits on 10 meters

The answer is B. The uplink 2 meter frequencies are in the range of 145.9 to 146.0 MHz and the 70 centimeter downlink frequencies are in the range of 435.1 to 435.2 MHz.

4BB-1B.4 What is <u>Mode L</u> in an amateur satellite?
A. Operation through a 70-centimeter receiver on a satellite that retransmits on 10 meters
B. Operation through a 23-centimeter receiver on a satellite that retransmits on 70 centimeters
C. Operation through a 70-centimeter receiver on a satellite that retransmits on 23 centimeters
D. Operation through a 10-meter receiver on a satellite that retransmits on 70 centimeters

The answer is B. The uplink band is from 1269.050 to 1269.850 MHz (23.6 cm) and the downlink band is from 436.950 to 436.150 MHz (69 cm).

4BB-1C.1 What is a <u>linear transponder?</u>
A. A repeater that passes only linear or CW signals
B. A device that receives and retransmits signals of any mode in a certain passband
C. An amplifier for SSB transmissions
D. A device used to change FM to SSB

The answer is B. A linear transponder will retransmit all types of signals on the transmitting frequency segment that it receives in the receiving frequency segment.

4BB-1C.2 What are the two basic types of <u>linear transponders</u> used in amateur satellites?
A. Inverting and non-inverting   B. Geostationary and elliptical
C. Phase 2 and Phase 3
D. Amplitude modulated and frequency modulated
    The answer is A. The transponder is the heart of the communications satellite. It receives a group of signals in one frequency spectrum and transmits the signals on another frequency spectrum. In some cases, it "inverts" the spectrum. By "inverting", we mean that a signal that is received at the low end of the band is retransmitted at the high end of the transmitting band, and vice versa. In other cases, the transponders are non-inverting, that is, a signal at the low end of the received band is retransmitted at the low end of the transmitted band.

4BB-1D.1 Why does the downlink frequency appear to vary by several kHz during a low-earth-orbit amateur satellite pass?
A. The distance between the satellite and ground station is changing, causing the Kepler effect
B. The distance between the satellite and ground station is changing, causing the Bernoulli effect
C. The distance between the satellite and ground station is changing, causing the Boyles' law effect
D. The distance between the satellite and ground station is changing, causing the Doppler effect
    The answer is D. When the distance between the signal source and the receiver changes, the frequency appears to change. This is true for sound, light or radio waves, and is known as the Doppler effect. This effect is commonly observed when a train with a whistle blowing, approaches a station. If you are at the station, the pitch of the train's whistle appears to increase as the train approaches. It appears to decrease as the train leaves.

4BB-1D.2 Why does the received signal from a Phase III amateur satellite exhibit a fairly rapid pulsed fading effect?
A. Because the satellite is rotating
B. Because of ionospheric absorption
C. Because of the satellite's low orbital altitude
D. Because of the Doppler effect
    The answer is A. Phase III amateur satellites stay in range for longer periods of time than Phase II satellites.

4BB-1D.3 What type of antenna can be used to minimize the effects of <u>spin modulation</u> and <u>Faraday rotation</u>?
A. A nonpolarized antenna
B. A circularly polarized antenna
C. An isotropic antenna
D. A log-periodic dipole array
    The answer is B. Faraday rotation refers to the changing polarization of a signal as it passes through the ionosphere.

4BB-2A.1 How often is a new frame transmitted in a fast-scan television system?
A. 30 times per second   B. 60 times per second
C. 90 times per second   D. 120 times per second
    The answer is A. Do not confuse frames with fields. There are 60 fields per second, but only 30 frames per second. There are two fields per frame. These rates were chosen so as to avoid flicker and power line interference.

4BB-2A.2 How many horizontal lines make up a fast-scan television frame?
A. 30   B. 60   C. 525   D. 1050
    The answer is C. Since there are two fields per frame, there are 262.5 lines per field.

4BB-2A.3 How is the interlace scanning pattern generated in a fast-scan television system?
A. By scanning the field from top to bottom
B. By scanning the field from bottom to top
C. By scanning even numbered lines in one field and odd numbered ones in the next
D. By scanning from left to right in one field and right to left in the next
    The answer is C. By scanning 262.5 lines in one field and the other 262.5 lines in the next field, we have the effect of 60 fields or "pictures" per second. This overcomes the flicker effect that would be apparent if we had 30 fields or "pictures" per second.

4BB-2A.4 What is blanking in a video signal?
A. Synchronization of the horizontal and vertical sync-pulses
B. Turning off the scanning beam while it is traveling from right to left and from bottom to top
C. Turning off the scanning beam at the conclusion of a transmission
D. Transmitting a black and white test pattern
    The answer is B. We don't want the beam to be visible when it is going from the right side of the picture tube back to the left side, and from the bottom of the tube back up to the top. If these lines were visible, the picture would be practically destroyed.

4BB-2A.5 What is the standard video voltage level between the sync tip and the whitest white at TV camera outputs and modulator inputs?
A. 1 volt peak-to-peak   B. 120 IEEE units
C. 12 volts dc   D. 5 volts RMS
    The answer is A. The 1 volt encompasses the entire video signal and the sync pulses. It does not include the span between the whitest white and the zero carrier.

4BB-2A.6 What is the bandwidth of a fast-scan television transmission?
A. 3 kHz   B. 10 kHz   C. 25 kHz   D. 6 MHz
    The answer is D. This is the standard bandwidth of a commercial TV signal. It includes all the upper sidebands, a small part of the

lower sidebands (vestigial sideband), and the sound.

4BB-2A.7 What is the standard video level, in percent PEV, for black?
A. 0%   B. 12.5%   C. 70%   D. 100%

The answer is C. PEV stands for Peak Envelope Voltage. Figure 4BB-2A.7, illustrates a graph of a line of picture information, including blanking and horizontal pulses. Note that the reference black level is 70% of the maximum carrier voltage.

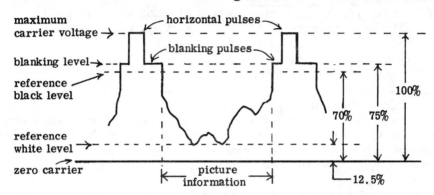

Fig. 4BB-2A.7. A line of picture information together with blanking information and horizontal pulses.

4BB-2A.8 What is the standard video level, in percent PEV, for white?
A. 0%   B. 12.5%   C. 70%   D. 100%

The answer is B. From Figure 4BB-2A.7, it can be seen that the white reference level is 12.5% of the maximum carrier voltage.

4BB-2A.9 What is the standard video level, in percent PEV, for blanking?
A. 0%   B. 12.5%   C. 75%   D. 100%

The answer is C. From Figure 4BB-2A.7, it can be seen that the blanking reference is 75% of the maximum carrier voltage.

SUBELEMENT 4BC
RADIO WAVE PROPAGATION
(2 questions)

4BC-1.1 What is the maximum separation between two stations communicating by moonbounce?
A. 500 miles maximum, if the moon is at perigee
B. 2,000 miles maximum, if the moon is at apogee
C. 5,000 miles maximum, if the moon is at perigee
D. Any distance as long as the stations have a mutual lunar window

The answer is D. Moonbounce is an advanced and somewhat experimental mode of VHF communication in which two stations communicate by their signals being reflected from the moon's surface. Because surface reflection from the moon is poor, signals are usually quite weak. Maximum power is required, along with high gain antennas, and low-noise receivers with high selectivity. Prearranged schedules are usually a requirement. This mode is often referred to as EME, or earth-moon-earth.

In order for two stations to communicate via the moon, they must both face the moon in such a way that each signal will strike the moon and return to earth, where it can be received by the other station.

4BC-1.2 What characterizes libration fading of an EME signal?
A. A slow change in the pitch of the CW signal
B. A fluttery, rapid irregular fading
C. A gradual loss of signal as the sun rises
D. The returning echo is several hertz lower in frequency than the transmitted signal

The answer is B. Libration fading is fading caused by the rocking motion of the moon as it travels in its orbit, and by the scattering of the signals from the irregular moon surface. This results in a rapid flutter of the signal received from the moon.

4BC-1.3 What are the best days to schedule EME contacts?
A. When the moon is at perigee   B. When the moon is full
C. When the moon is at apogee
D. When the weather at both stations is clear

The answer is A. Perigee is the point in the moon's orbit when it is closest to the earth. Since distance is a factor in receiving E-M-E signals, it is important to schedule E-M-E contacts when the moon is closest to the earth.

4BC-1.4 What type of receiving system is required for EME communications?
A. Equipment capable of reception on 14 MHz
B. Equipment with very low dynamic range
C. Equipment with very low gain
D. Equipment with very low noise figures

The answer is D. The antenna must be a large, high gain antenna. The receiver must have a low noise figure, high gain and excellent selectivity. See answer 4BC-1.1.

4BC-1.5 What type of transmitting system is required for EME communications?

A. A transmitting system capable of operation on the 21 MHz band
B. A transmitting system capable of producing a very high ERP
C. A transmitting system using an unmodulated carrier
D. A transmitting system with a high second harmonic output

The answer is B. The transmitting signal must be stable and the power must be the maximum legal amount because of the weak signals that result from the moon's reflection. The antenna must have high gain.

4BC-2.1 When the earth's atmosphere is struck by a meteor, a cylindrical region of free electrons is formed at what layer of the ionosphere?

A. The F1 layer B. The E layer C. The F2 layer D. The D layer

The answer is B. At certain times of the year, the earth is regularly bombarded by "meteor showers", sometimes lasting for hours. This produces spectacular visual displays, accompanied by prolonged periods of ionization in the E layer, during which time, long distance communication becomes possible. This effect, also called meteor scatter, is most noticeable on the 6-meter band and, to a lesser extent, on 2 meters.

4BC-2.2 Which range of frequencies is well suited for meteor- scatter communications?

A. 1.8 - 1.9 MHz
B. 10 - 14 MHz
C. 28 - 148 MHz
D. 220 - 450 MHz

The answer is C. The 50-54 MHz band is well suited for meteor-burst transmissions.

4BC-3.1 What is transequatorial propagation?

A. Propagation between two points at approximately the same distance north and south of the magnetic equator
B. Propagation between two points on the magnetic equator
C. Propagation between two continents by way of ducts along the magnetic equator
D. Propagation between any two stations at the same latitude

The answer is A. Trans-equatorial propagation or trans-equatorial scatter, is found on the six meter band. It occurs between stations located on opposite sides of the equator and whose signals travel in a north-south direction. It occurs mostly during peak periods of sun spot activity, and is generally an afternoon and evening phenomena.

4BC-3.2 What is the maximum range for signals using transequatorial propagation?

A. About 1,000 miles
B. About 2,500 miles
C. About 5,000 miles
D. About 7,500 miles

The answer is C. Ionization is quite strong at the equator, and trans-equatorial scatter can provide communications up to over 5,000 miles.

4BC-3.3 What is the best time of day for <u>transequatorial propagation</u>?
A. Morning   B. Noon
C. Afternoon or early evening
D. Trans-equatorial propagation only works at night
     The answer is C. See answer 4BC-3.1.

4BC-4.1 If a beam antenna must be pointed in a direction 180 degrees away from a station to receive the strongest signals, what type of propagation is probably occurring?
A. Transequatorial propagation
B. Sporadic-E propagation
C. Long-path propagation
D. Auroral propagation
     The answer is C. The shortest path between two stations usually provides the best communication between the two stations. For instance, if a U.S. station is in contact with a French station, the antenna of the U.S. station will be pointed East, toward France. The French station's antenna will be pointed West, toward the U.S. This is the shortest direct path. There are times, however, when propagation conditions are poor along this short path. If the two stations rotate their antennas 180 degrees, propagation conditions may improve. Their signals are now travelling along the LONG PATH, across Asia. This type of propagation is referred to as LONG-PATH PROPAGATION.

4BC-5.1 What is the name for a type of propagation in which radio signals travel along the <u>terminator,</u> which separates daylight from darkness?
A. Transequatorial propagation
B. Sporadic-E propagation
C. Long-path propagation
D. Gray-line propagation
     The answer is D. At any given time, that part of the earth that faces the sun is illuminated (daylight) and the other part of the earth is in darkness. The region or band surrounding the earth at the line where lightness and darkness meet is called the "gray-line" or "terminator". Radio propagation along this "gray line" is called "gray-line propagation".

4BD-1A.1 How does a <u>spectrum analyzer</u> differ from a conventional time-domain oscilloscope?

A. The oscilloscope is used to display electrical signals while the spectrum analyzer is used to measure ionospheric reflection

B. The oscilloscope is used to display electrical signals in the frequency domain while the spectrum analyzer is used to display electrical signals in the time domain

C. The oscilloscope is used to display electrical signals in the time domain while the spectrum analyzer is used to display electrical signals in the frequency domain

D. The oscilloscope is used for displaying audio frequencies and the spectrum analyzer is used for displaying radio frequencies

The answer is C. When we speak about time domain, we mean that the horizontal distance on the oscilloscope is a function of time. If we examine a signal on an oscilloscope, that part of the signal on the right of the screen occurs later IN TIME than the part of the signal on the left of the screen. When we speak about frequency domain, we mean that the horizontal distance on the screen is a function of the frequency. When we view a spectrum analyzer, the signals on the right side of the screen occur at different frequencies than those on the left side of the screen. See answer 4BD-1B.1 for a discussion of a spectrum analyzer.

4BD-1A.2 What does the horizontal axis of a <u>spectrum analyzer</u> display?

A. Amplitude   B. Voltage   C. Resonance   D. Frequency

The answer is D. See answer 4BD-1A.1.

4BD-1A.3 What does the vertical axis of a <u>spectrum analyzer</u> display?

A. Amplitude   B. Duration   C. Frequency   D. Time

The answer is A. The vertical axis of a spectrum analyzer, or an oscilloscope, is a function of amplitude. For instance, the signal at the center of the screen of Figure 4BD-1B.1 is much greater than the two signals at the ends of the screen.

4BD-1B.1 What test instrument can be used to display spurious signals in the output of a radio transmitter?

A. A spectrum analyzer   B. A wattmeter
C. A logic analyzer   D. A time-domain reflectometer

The answer is A. A spectrum analyzer is a combination of an oscilloscope and a receiver. A given bandwidth from the receiver portion is displayed on the CRT screen, allowing observation of the input signal and the frequency space above and below the input signal. Spectrum analyzers are especially useful in determining if any spurious emissions are present in a transmitter's output. See

Figure 4BD-1B.1. It shows the transmitter's signal, together with spurious output signals, above and below the transmitter signal's frequency.

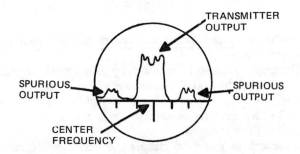

4BD-1B.1 Spectrum analyzer display.

4BD-1B.2 What test instrument is used to display intermodulation distortion products from an SSB transmitter?

A. A wattmeter   B. A spectrum analyzer
C. A logic analyzer
D. A time-domain reflectometer

The answer is B. Transmitter intermodulation distortion is the spurious, unwanted signals that come about from the mixing of signals in a non-linear stage. The two-tone test is the standard method used to determine the extent of the intermodulation distortion.

4BD-2A.1 What advantage does a logic probe have over a voltmeter for monitoring logic states in a circuit?

A. A logic probe has fewer leads to connect to a circuit than a voltmeter
B. A logic probe can be used to test analog and digital circuits
C. A logic probe can be powered by commercial ac lines
D. A logic probe is smaller and shows a simplified readout

The answer is D. There are only two logic states – 0 and 1, or "off" and "on". It isn't necessary to use a voltmeter to monitor the logic states in a circuit. A simple logic probe that indicates "on" and "off" will suffice.

4BD-2A.2 What piece of test equipment can be used to directly indicate high and low logic states?

A. A galvanometer   B. An electroscope
C. A logic probe   D. A Wheatstone bridge

The answer is C. A logic probe is a test instrument that indicates the logic level of the point that it is applied to. Indication is by a light (usually a LED) that lights when the test point is 1 (high), and goes out when the test point is 0 (low). A test point with a pulsing signal (one that is rapidly switching back and forth between 0 and 1) will result in a rapidly flickering light. One disadvantage of the logic probe is that many logic circuits need to be tested at several points simultaneously, while a logic probe can only test one point at a time.

4BD-2A.3 What is a logic probe used to indicate?
A. A short-circuit fault in a digital-logic circuit
B. An open-circuit failure in a digital-logic circuit
C. A high-impedance ground loop
D. High and low logic states in a digital-logic circuit
    The answer is D. See answer 4BD-2A.2.

4BD-2B.1 What piece of test equipment besides an oscilloscope can be used to indicate pulse conditions in a digital-logic circuit?
A. A logic probe   B. A galvanometer
C. An electroscope   D. A Wheatstone bridge
    The answer is A. See answer 4BD-2A.2.

4BD-3A.1 What is one of the most significant problems you might encounter when you try to receive signals with a mobile station?
A. Ignition noise   B. Doppler shift
C. Radar interference   D. Mechanical vibrations
    The answer is A. Ignition noise comes from sparking, which produces a popping sound that increases in frequency, but not in pitch, as the engine speed increases.

4BD-3A.2 What is the proper procedure for suppressing electrical noise in a mobile station?
A. Apply shielding and filtering where necessary
B. Insulate all plane sheet metal surfaces from each other
C. Apply antistatic spray liberally to all non-metallic surfaces
D. Install filter capacitors in series with all dc wiring
    The answer is A. Some steps to reduce electrical noise in a mobile station are: (1) Use resistor-type spark plugs or resistance-type ignition wiring. (2) Install a bypass capacitor from the primary of the ignition coil to ground. (3) Install a bypass capacitor across the moving contacts of the voltage regulator. (4) In extreme cases, the entire ignition system must be shielded, using special kits.

4BD-3A.3 How can ferrite beads be used to suppress ignition noise?
A. Install them in the resistive high voltage cable every 2 years
B. Install them between the starter solenoid and the starter motor
C. Install them in the primary and secondary ignition leads
D. Install them in the antenna lead to the radio
    The answer is C. A ferrite bead is a small cylinder-shaped object made of iron-type material that slips over a wire conductor. It increases the inductance at the point where it slips on the wire and forms an effective RF choke. This helps to suppress harmonics and other high frequency interference. Placing the beads on the primary and secondary ignition leads is an important step in suppressing ignition noise.

4BD-3A.4 How can ensuring good electrical contact between connecting metal surfaces in a vehicle reduce spark plug noise?
A. It reduces the spark gap distance, causing a lower frequency spark
B. It helps radiate the spark plug noise away from the vehicle
C. It reduces static buildup on the vehicle body
D. It encourages lower frequency electrical resonances in the vehicle

The answer is D. Shielding the engine compartment confines the interference to the engine compartment and doesn't let it get to the receiver. Shielding absorbs the interfering signals and converts it to heat, rather than allowing it to radiate to the receiver.

The various metal surfaces in a car should be bonded together to form one large ground. Otherwise, they tend to reradiate the interfering signals. Metal surfaces that are not bonded together also tend to produce static.

4BD-3B.1 How can alternator whine be minimized?
A. By connecting the radio's power leads to the battery by the longest possible path
B. By connecting the radio's power leads to the battery by the shortest possible path
C. By installing a high pass filter in series with the radio's dc power lead to the vehicle's electrical system
D. By installing filter capacitors in series with the dc power lead

The answer is B. Alternator interference shows up as a whining sound that changes pitch as the engine speed is changed. In addition to answer B, alternator interference can be reduced by installing special coaxial capacitors at the alternator. The capacitor case should be grounded to the alternator frame.

4BD-3B.2 How can conducted and radiated noise caused by an automobile alternator be suppressed?
A. By installing filter capacitors in series with the dc power lead and by installing a blocking capacitor in the field lead
B. By connecting the radio's power leads to the battery by the longest possible path and by installing a blocking capacitor in series with the positive lead
C. By installing a high pass filter in series with the radio's power lead to the vehicle's electrical system and by installing a low-pass filter in parallel with the field lead
D. By connecting the radio power leads directly to the battery and by installing coaxial capacitors in the alternator leads

The answer is D. See answer 4BD-3B.1.

4BD-3C.1 What is a major cause of atmospheric static?
A. Sunspots   B. Thunderstorms
C. Airplanes   D. Meteor showers

The answer is B. Static is caused by lightning and similar natural electrical disturbances.

4BD-3D.1 How can you determine if a line-noise interference problem is being generated within your home?
A. Check the power-line voltage with a time-domain reflectometer
B. Observe the ac waveform on an oscilloscope
C. Turn off the main circuit breaker and listen on a battery- operated radio
D. Observe the power-line voltage on a spectrum analyzer

The answer is C. If the noise is still present when the receiver is battery operated and disconnected from the power line, then we know that the noise is not generated in the home. If the noise disappears when the receiver is disconnected from the power lines, then we know that the noise is generated in the home and is fed to the receiver via the power lines.

4BD–4.1 What is the main drawback of a wire–loop antenna for direction finding?

A. It has a bidirectional pattern broadside to the loop

B. It is non–rotatable

C. It receives equally well in all directions

D. It is practical for use only on VHF bands

The answer is A. A loop antenna can be used for direction–finding purposes. A loop antenna is actually a coil whose shape may be circular or rectangular. Figure 4BD–4.1A illustrates a square loop. The loop has maximum reception when its plane is parallel to the plane of the signal propagation. This is shown in Figure 4BD–4.1B.
This pattern is referred to as bidirectional or figure 8. The loop antenna receives minimum signal strength when its plane faces the oncoming signal. This is referred to as a NULL position.

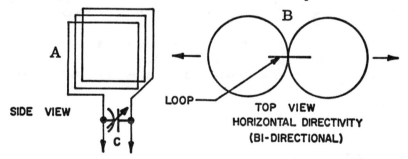

Figure 4BD-4.1. A loop antenna with its horizontal pattern.

4BD–4.2 What directional pattern is desirable for a direction– finding antenna?

A. A non–cardioid pattern

B. Good front–to–back and front–to–side ratios

C. Good top–to–bottom and front–to–side ratios

D. Shallow nulls

The answer is B. We would like a sharp maximum or minimum point when we rotate the loop. In practice, however, the null point is sharp while the maximum point is broad. So we tune for null.

4BD–4.3 What is the triangulation method of direction finding?

A. Using the geometric angle of ground waves and sky waves emanating from the same source to locate the signal source

B. A fixed receiving station uses three beam headings to plot the signal source on a map

C. Beam headings from several receiving locations are used to plot the signal source on a map

D. The use of three vertical antennas to indicate the location of the signal source

The answer is C. To use a loop for direction-finding, the loop is rotated for maximum or null. With the aid of a compass and a map, a line is plotted along the path of maximum pickup. The loop is taken to a second location and another line is plotted. The point where the two lines intersect is the location of the station. Figure 4BD-4.3 illustrates the triangulation method of direction finding.

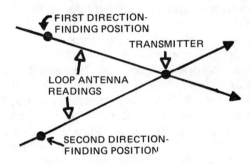

Figure 4BD-4.3. Triangulation method of direction finding.

4BD-4.4 Why is an RF attenuator desirable in a receiver used for direction finding?
A. It narrows the bandwidth of the received signal
B. It eliminates the effects of isotropic radiation
C. It reduces loss of received signals caused by antenna pattern nulls
D. It prevents receiver overload from extremely strong signals

The answer is D. If the receiver overloads from strong signals, it is impossible to get sharp nulls or good maximums.

4BD-4.5 What is a <u>sense antenna?</u>
A. A vertical antenna added to a loop antenna to produce a cardioid reception pattern
B. A horizontal antenna added to a loop antenna to produce a cardioid reception pattern
C. A vertical antenna added to an Adcock antenna to produce an omnidirectional reception pattern
D. A horizontal antenna added to an Adcock antenna to produce a cardioid reception pattern

The answer is A. A loop antenna, by itself, gives us a figure 8 or bidirectional pattern. By adding a vertical antenna (called a "sense" antenna) to the loop, we change the pattern into a unidirectional pattern. An actual graph of the pattern resembles a heart, hence the name, cardioid. Figure 4BD-4.5 illustrates a cardioid pattern.

4BD-4.6 What type of antenna is most useful for sky-wave reception in radio direction finding?
A. A log-periodic dipole array
B. An isotropic antenna
C. A circularly polarized antenna

D. An Adcock antenna

The answer is D. An Adcock antenna consists of two large vertical elements, connected together at their center by a transmission line. In direction finding, loop antennas are good for low frequency, ground wave reception, whereas Adcock antennas are good for higher frequency sky wave reception.

Figure 4BD-4.5. A cardioid pattern.

MAXIMUM RADIATION

4BD-4.7 What is a <u>loop antenna</u>?
A. A circularly polarized antenna
B. A coil of wire used as an antenna in FM broadcast receivers
C. A wire loop used in radio direction finding
D. An antenna coupled to the feed line through an inductive loop of wire

The answer is C. See answer 4BD-4.1.

4BD-4.8 How can the output voltage of a loop antenna be increased?
A. By reducing the permeability of the loop shield
B. By increasing the number of wire turns in the loop while reducing the area of the loop structure
C. By reducing either the number of wire turns in the loop, or the area of the loop structure
D. By increasing either the number of wire turns in the loop, or the area of the loop structure

The answer is D.

4BD-4.9 Why is an antenna system with a cardioid pattern desirable for a direction-finding system?
A. The broad side responses of the cardioid pattern can be aimed at the desired station
B. The deep null of the cardioid pattern can pinpoint the direction of the desired station
C. The sharp peak response of the cardioid pattern can pinpoint the direction of the desired station
D. The high radiation angle of the cardioid pattern is useful for short-distance direction finding

The answer is B. See answers 4BD-4.1 and 4BD-4.5.

4BD-4.10 What type of terrain can cause errors in direction finding?
A. Homogeneous terrain   B. Smooth grassy terrain
C. Varied terrain   D. Terrain with no buildings or mountains

The answer is C. A varied terrain causes reflections of the signal and multipath effects, which will introduce errors in direction finding.

4BE-1.1 What is the <u>photoconductive effect</u>?
A. The conversion of photon energy to electromotive energy
B. The increased conductivity of an illuminated semiconductor junction
C. The conversion of electromotive energy to photon energy
D. The decreased conductivity of an illuminated semiconductor junction
The answer is B. The photoconductive effect refers to the change in the resistance (and therefore, the current) in certain semiconductors when exposed to different amounts of light. If we go from "no light" to "light", or if we increase the amount of light, the resistance will decrease and the current will increase.

4BE-1.2 What happens to photoconductive material when light shines on it?
A. The conductivity of the material increases
B. The conductivity of the material decreases
C. The conductivity of the material stays the same
D. The conductivity of the material becomes temperature dependent
The answer is A. See answer 4BE-1.1.

4BE-1.3 What happens to the resistance of a photoconductive material when light shines on it?
A. It increases   B. It becomes temperature dependent
C. It stays the same   D. It decreases
The answer is D. See answer 4BE-1.1.

4BE-1.4 What happens to the conductivity of a semiconductor junction when it is illuminated?
A. It stays the same   B. It becomes temperature dependent
C. It increases   D. It decreases
The answer is C. Conductivity is the opposite of resistance. See answer 4BE-1.1.

4BE-1.5 What is an <u>optocoupler</u>?
A. A resistor and a capacitor
B. A frequency modulated helium-neon laser
C. An amplitude modulated helium-neon laser
D. An LED and a phototransistor
The answer is D. An optocoupler is the same as an optoisolator. It consists of an LED and a phototransistor in one package. Optocouplers can be used for line voltage regulation.

4BE-1.6 What is an <u>optoisolator</u>?
A. An LED and a phototransistor
B. A P-N junction that develops an excess positive charge when exposed to light
C. An LED and a capacitor

D. An LED and a solar cell
   The answer is A. See answer 4BE-1.5.

4BE-1.7 What is an <u>optical shaft encoder</u>?
A. An array of optocouplers chopped by a stationary wheel
B. An array of optocouplers whose light transmission path is controlled
   by a rotating wheel
C. An array of optocouplers whose propagation velocity is controlled
   by a stationary wheel
D. An array of optocouplers whose propagation velocity is controlled
   by a rotating wheel
   The answer is B.

4BE-1.8 What does the <u>photoconductive effect</u> in crystalline solids produce a noticeable change in?
A. The capacitance of the solid   B. The inductance of the solid
C. The specific gravity of the solid
D. The resistance of the solid
   The answer is D. This, in turn, causes a current change.

4BE-2A.1 What is the meaning of the term <u>time constant</u> of an RC circuit?
A. The time required to charge the capacitor in the circuit to 36.8%
   of the supply voltage
B. The time required to charge the capacitor in the circuit to 36.8%
   of the supply current
C. The time required to charge the capacitor in the circuit to 63.2%
   of the supply current
D. The time required to charge the capacitor in the circuit to 63.2%
   of the supply voltage
   The answer is D. When a supply voltage is placed across a resistor
and capacitor in series, as in Figure 4BE-2A.1A, the capacitor does
not charge up to the supply voltage immediately. As the first electrons
from the voltage source appear on the capacitor, the capacitor builds
up a counter-emf to the source voltage, and it takes a certain amount
of time until the voltage across the capacitor is the same as the supply
voltage. The term "time constant" refers to the amount of time it takes
for the voltage on the capacitor to reach 63.2% of the supply voltage.

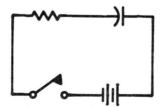

4BE-2A.1A. A resistor and capacitor in series.

   Curve A of Figure 4BE-2A.1B is a graph of the voltage build-up
on the capacitor. The horizontal line represents time-constants. The
vertical line represents the percent of full voltage or supply voltage.

Notice that at the end of one time constant, curve A reaches 63.2% of the full voltage.

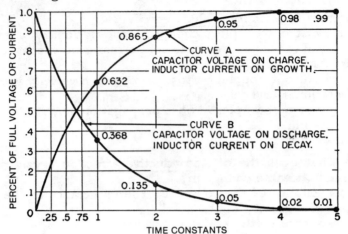

4BE-2A.1B. Universal time-constant curves.

4BE-2A.2 What is the meaning of the term <u>time constant</u> of an RL circuit?

A. The time required for the current in the circuit to build up to 36.8% of the maximum value

B. The time required for the voltage in the circuit to build up to 63.2% of the maximum value

C. The time required for the current in the circuit to build up to 63.2% of the maximum value

D. The time required for the voltage in the circuit to build up to 36.8% of the maximum value

The answer is C. When a coil and a resistor in series are hooked up to a source of current, the final value of current does not immediately appear in the coil. The first electrons that flow through the coil cause a magnetic field build-up around the coil, which induces an emf in the coil. This emf opposes the original supply current, and it takes a certain amount of time for the current in the coil to reach its final value. The time constant of an RL circuit refers to the time it takes for the current in the coil to build up to 63.2% of its final value.

Curve A of Figure 4BE-2A.1B is a graph of the current build up in the coil, in addition to being the graph of the voltage build up across a capacitor. This is why we refer to the curves of Figure 4BE-2A.1B as UNIVERSAL TIME-CONSTANT curves.

4BE-2A.3 What is the term for the time required for the capacitor in an RC circuit to be charged to 63.2% of the supply voltage?

A. An exponential rate of one   B. One time constant

C. One exponential period   D. A time factor of one

The answer is B. See answer 4BE-2A.1.

4BE-2A.4 What is the term for the time required for the current in

an RL circuit to build up to 63.2% of the maximum value?
A. One time constant   B. An exponential period of one
C. A time factor of one   D. One exponential rate
    The answer is A. See answer 4BE-2A.2.

4BE-2A.5 What is the term for the time it takes for a charged capacitor in an RC circuit to discharge to 36.8% of its initial value of stored charge?
A. One discharge period
B. An exponential discharge rate of one
C. A discharge factor of one
D. One time constant
    The answer is D. Just as it takes time for a capacitor to charge up, is takes time for the capacitor to discharge. The time constant in seconds or microseconds is the time it takes for the capacitor to discharge to 36.8% of its initial value. See curve B of Fig. 4BE-2A.1B.

4BE-2A.6 What is meant by back EMF?
A. A current equal to the applied EMF
B. An opposing EMF equal to R times C (RC) percent of the applied EMF
C. A current that opposes the applied EMF
D. A voltage that opposes the applied EMF
    The answer is D. A back-emf is the same as a counter-emf. It is a voltage that opposes the applied voltage from the source. For example, a coil develops a back-emf that opposes the applied emf. This particular back-emf gives us the inductive reactance.

4BE-2B.1 After two time constants, the capacitor in an RC circuit is charged to what percentage of the supply voltage?
A. 36.8%   B. 63.2%   C. 86.5%   D. 95%
    The answer is C. At the end of the first time constant, it charges up to 63.2% of the applied voltage. By the end of the second time constant, it charges to 63.2% of the balance. The balance is 100% – 63.2% or 36.8%. Therefore, in the second time constant, it charges to 63.2% of 36.8% or 23.26%. This is arrived at by simply multiplying the percentages as follows:

$$63.2\% \text{ of } 36.8\% = .632 \times .368 = .2326 \text{ which is } 23.26\%$$

The total percentage for the two time constants is 63.2% + 23.26%, which is equal to 86.46%.
    This can be verified by examining curve A of Figure 4BE-2A.1B. Note that, at the end of the second time-constant, curve A is at 86.46% of the full value.

4BE-2B.2 After two time constants, the capacitor in an RC circuit is discharged to what percentage of the starting voltage?
A. 86.5%   B. 63.2%   C. 36.8%   D. 13.5%
    The answer is D. The reasoning here is similar to, but somewhat different than answer 4BE-2B.1. At the end of the first time constant,

it discharges to 36.8% of its initial value. This means it still has 36.8% of the initial value. During the second time constant, it discharges to 36.8% of the balance of 36.8%. The balance remaining after two time constants is, therefore, 36.8% of 36.8%, which is 13.54%. This is arrived at by simply multiplying percentages:

36.8% of 36.8% = .368 x .368 = .1354 which is 13.54%

This can be verified by examining curve B of Figure 4BE-2A.1B. Note that, at the end of the second time-constant, the voltage across the capacitor has fallen to 13.54% of the full voltage.

4BE-2B.3 What is the time constant of a circuit having a 100- microfarad capacitor in series with a 470-kilohm resistor?
A. 4700 seconds   B. 470 seconds   C. 47 seconds   D. 0.47 seconds
The answer is C. The time constant, in seconds, is equal to the product of the capacity, in farads, and the resistance, in ohms. We put this in the form of an equation and solve the problem.

Time constant = CR = .0001 X 470,000 = 47 seconds

Note that the capacity is given in microfarads and must be changed to farads. The resistance is given in kilohms and must be changed to ohms.

4BE-2B.4 What is the time constant of a circuit having a 220- microfarad capacitor in parallel with a 1-megohm resistor?
A. 220 seconds   B. 22 seconds   C. 2.2 seconds   D. 0.22 seconds
The answer is A. We arrive at this answer by using the time constant formula of answer 4BE-2B.3.

TC = CR = .000220 X 1,000,000 = 220 seconds

Figure 4BE-2B.4 illustrates a capacitor and a resistor in parallel across a source. When we speak about the time constant of a parallel circuit we are concerned primarily about the discharge of the capacitor through the resistor when the source is removed. In a sense, this action behaves like a series circuit.

Technically speaking, we have an almost zero (0) time constant on charge because there is no resistance between the capacitor and the source (as there is in a series circuit). The capacitor charges up to full value instantly.

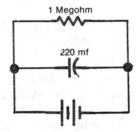

Fig. 4BE-2B.4. A resistor and capacitor in parallel.

4BE-2B.5 What is the time constant of a circuit having two 100-microfarad capacitors and two 470-kilohm resistors all in series?
A. 470 seconds   B. 47 seconds   C. 4.7 seconds   D. 0.47 seconds
The answer is B. We must first find the total capacitance in farads.

$$C_T = \frac{100 \times 100}{100 + 100} = \frac{10,000}{200} = 50 \text{ mfd} = .00005 \text{ fd.}$$

We then find the total resistance in ohms.

$$470K + 470K = 940K = 940,000 \text{ ohms.}$$

We then substitute these values in the Time Constant formula.

$$TC = CR = .00005 \times 940,000 = 47 \text{ seconds.}$$

4BE-2B.6 What is the time constant of a circuit having two 100-microfarad capacitors and two 470-kilohm resistors all in parallel?
A. 470 seconds   B. 47 seconds   C. 4.7 seconds   D. 0.47 seconds
The answer is B. We first find the total equivalent capacitance. In a parallel circuit, it is the sum of the two capacitors. 100 mf + 100 mf = 200 mf. We then find the total resistance of the two parallel resistors.

$$R_{total} = \frac{R1 \times R2}{R1 + R2} = \frac{470K \times 470K}{470K + 470K} = \frac{220,900K}{940K} = 235 \text{ Kohms}$$

We then find the Time Constant of the circuit.

$$TC = CR = 235,000 \times .000200 = 47 \text{ seconds}$$

See discussion in answer 4BE-2B.4.

4BE-2B.7 What is the time constant of a circuit having two 220-microfarad capacitors and two 1-megohm resistors all in series?
A. 55 seconds   B. 110 seconds   C. 220 seconds   D. 440 seconds
The answer is C. We use the method in answer 4BE-2B.5 to solve this problem.

$$C_{total} = \frac{220 \times 220}{220 + 220} = \frac{48,400}{440} = 110 \text{ mfd} = .00011 \text{ fd.}$$

$$R_{total} = 1 \text{ megohm} + 1 \text{ meghom} = 2 \text{ megohms} = 2,000,000 \text{ ohms.}$$

$$TC = CR = .00011 \times 2,000,000 = 220 \text{ seconds.}$$

4BE-2B.8 What is the time constant of a circuit having two 220-microfarad capacitors and two 1-megohm resistors all in parallel?
A. 22 seconds   B. 44 seconds   C. 220 seconds   D. 440 seconds
The answer is C. This problem is solved in exactly the same manner as 4BE-2B.6. The total capacitance is equal to 220 mf + 220 mf = 440 mf. The total resistance is equal to:

$$R_{total} = \frac{1\text{Meg} \times 1\text{Meg}}{1\text{Meg} + 1\text{Meg}} = \frac{1\text{Meg}}{2\text{Meg}} = .5 \text{ Megohm}$$

We then substitute these values in the time constant formula to find the time constant.

$$TC = CR = .000440 \times 500,000 = 220 \text{ seconds}$$

See discussion in answer 4BE-2B.4

4BE-2B.9 What is the time constant of a circuit having one 100-microfarad capacitor, one 220-microfarad capacitor, one 470- kilohm resistor and one 1-megohm resistor all in series?
A. 68.8 seconds B. 101.1 seconds C. 220.0 seconds D. 470.0 seconds
The answer is B. We use the method in answer 4BE-2B.5 to solve this problem.

$$C_{total} = \frac{100 \times 220}{100 + 220} = \frac{22,000}{320} = 68.75 \text{ mfd} = .00006875 \text{ fd.}$$

$R_{total} = 470\text{Kohms} + 1 \text{ megohm} = 1,470,000 \text{ ohms.}$

$TC = CR = .00006875 \times 1,470,000 = 101 \text{ seconds.}$

4BE-2B.10 What is the time constant of a circuit having a 470-microfarad capacitor and a 1-megohm resistor in parallel?
A. 0.47 seconds B. 47 seconds   C. 220 seconds   D. 470 seconds
The answer is D. We arrive at this answer by using the time constant formula.

$$TC = CR = .000470 \times 1,000,000 = 470 \text{ seconds}$$

See discussion in answer 4BE-2B.4.

4BE-2B.11 What is the time constant of a circuit having a 470-microfarad capacitor in series with a 470-kilohm resistor?
A. 221 seconds   B. 221000 seconds
C. 470 seconds   D. 470000 seconds
The answer is A. After converting microfarads to farads and kilohms to ohms, we use the Time Constant formula to solve the problem.

$$470 \text{ mf} = .00047 \text{ f} \quad 470 \text{ kilohm} = 470,000 \text{ ohms}$$

$$TC = CR = .00047 \times 470,000 = 220.9 \text{ seconds}$$

4BE-2B.12 What is the time constant of a circuit having a 220-microfarad capacitor in series with a 470-kilohm resistor?
A. 103 seconds   B. 220 seconds   C. 470 seconds   D. 470000 seconds
The answer is A. After converting microfarads to farads and kilohms to ohms, we use the Time Constant formula to solve the problem.

220 mf = .00022 f   470 kilohm = 470,000 ohms

TC = CR = .00022 X 470,000 = 103.4 seconds

4BE-2B.13 How long does it take for an initial charge of 20 V dc to decrease to 7.36 V dc in a 0.01-microfarad capacitor when a 2- megohm resistor is connected across it?
A. 12.64 seconds B. 0.02 seconds C. 1 second   D. 7.98 seconds
The answer is B. After converting microfarads to farads and megohms to ohms, we use the Time Constant formula to solve the problem.

.01 mf = .00000001 f   2 megohms = 2,000,000 ohms
TC = CR = .00000001 X 2,000,000 = .02 seconds

Since it takes one Time Constant of time to drop to 36.8% of 20 volts or 7.36 volts (.368 X 20 = 7.36), and since one Time Constant is .02 seconds, then we know that it takes .02 seconds to drop to 7.36 volts.

4BE-2B.14 How long does it take for an initial charge of 20 V dc to decrease to 2.71 V dc in a 0.01-microfarad capacitor when a 2- megohm resistor is connected across it?
A. 0.04 seconds B. 0.02 seconds C. 7.36 seconds D. 12.64 seconds
The answer is A. From the answer of 4BE-2B.13, we know that the Time Constant is 0.02 seconds, and we know that after one Time Constant, the voltage on the capacitor is 7.36 volts. After the second Time Constant of elapsed time, the voltage becomes 7.36 X 36.8% or 2.71 volts. Two time constants is equal to 2 X .02 = .04 seconds. Therefore, the total time for the capacitor to get down to 2.71 volts is .04 seconds.

4BE-2B.15 How long does it take for an initial charge of 20 V dc to decrease to 1 V dc in a 0.01-microfarad capacitor when a 2- megohm resistor is connected across it?
A. 0.01 seconds B. 0.02 seconds C. 0.04 seconds D. 0.06 seconds
The answer is D. From the answer of 4BE-2B.13, we know that the time constant is 0.02 seconds. We also know from answer 4BE-2B.14 that after two Time Constants (.04 seconds), the voltage across the capacitor falls to 2.71 volts. In the next Time Constant of .02 seconds, the voltage will fall to 36.8% of 2.71 volts or 1 volt. (.368 X 2.71 = 1). It therefore takes .06 seconds (.04 + .02) or three Time Constants to reach 1 volt.

4BE-2B.16 How long does it take for an initial charge of 20 V dc to decrease to 0.37 V dc in a 0.01-microfarad capacitor when a 2- megohm resistor is connected across it?
A. 0.08 seconds B. 0.6 seconds   C. 0.4 seconds   D. 0.2 seconds
The answer is A. See answers 4BE-2B.13 through 4BE-2B.15. Using the same reasoning as in answer 4BE-2B.15, an additional Time Constant of .02 seconds will bring the voltage across the capacitor down to 0.37 volts (1 volt X .368 = .37 volts). The total time is, therefore, .08 seconds (.06 + .02 = .08).

4BE-2B.17 How long does it take for an initial charge of 20 V dc to decrease to 0.13 V dc in a 0.01-microfarad capacitor when a 2-megohm resistor is connected across it?
A. 0.06 seconds   B. 0.08 seconds   C. 0.1 seconds   D. 1.2 seconds

The answer is C. Using the same reasoning as in the previous questions (4BE-2B.13 through 4BE-2B.16), an additional Time Constant of .02 seconds will bring the voltage across the capacitor down to .13 volts (.37 volts X .368 = .1362 volts). The total time is, therefore, 0.1 seconds (.08 + .02 = .1).

4BE-2B.18 How long does it take for an initial charge of 800 V dc to decrease to 294 V dc in a 450-microfarad capacitor when a 1-megohm resistor is connected across it?
A. 80 seconds   B. 294 seconds   C. 368 seconds   D. 450 seconds
The answer is D. The Time Constant is

$$TC = CR = .00045 \times 1,000,000 = 450 \text{ seconds.}$$

In one Time Constant, the voltage across the capacitor will drop to 36.8% of its initial voltage.

$$800 \text{ volts} \times .368 = 294 \text{ volts.}$$

It will, therefore, take one Time Constant or 450 seconds to drop to 294 volts.

4BE-2B.19 How long does it take for an initial charge of 800 V dc to decrease to 108 V dc in a 450-microfarad capacitor when a 1-megohm resistor is connected across it?
A. 225 seconds   B. 294 seconds   C. 450 seconds   D. 900 seconds

The answer is D. See answer 4BE-2B.18. In the second Time Constant, the voltage will fall to 36.8% of what it was at the end of the first Time Constant.

$$294 \text{ volts} \times .368 = 108 \text{ volts.}$$

It will, therefore, take two Time Constants or 2 X 450 seconds (900 seconds) to fall to 108 volts.

4BE-2B.20 How long does it take for an initial charge of 800 V dc to decrease to 39.9 V dc in a 450-microfarad capacitor when a 1-megohm resistor is connected across it?
A. 1350 seconds   B. 900 seconds   C. 450 seconds   D. 225 seconds

The answer is A. See answers 4BE-2B.18 and 4BE-2B.19. In the third Time Constant, the voltage will fall to 36.8% of 108 volts.

$$108 \text{ volts} \times .368 = 39.7 \text{ volts.}$$

It will, therefore, take approximately three Time Constants, or 3 X 450 seconds (1350 seconds) to fall to 39.9 volts.

4BE-2B.21 How long does it take for an initial charge of 800 V dc to decrease to 40.2 V dc in a 450-microfarad capacitor when a 1-megohm resistor is connected across it?
A. Approximately 225 seconds   B. Approximately 450 seconds

C. Approximately 900 seconds   D. Approximately 1350 seconds

The answer is D. See answer 4BE-2B.20. This question is almost the same as 4BE-2B.20.

4BE-2B.22 How long does it take for an initial charge of 800 V dc to decrease to 14.8 V dc in a 450-microfarad capacitor when a 1-megohm resistor is connected across it?
A. Approximately 900 seconds   B. Approximately 1350 seconds
C. Approximately 1804 seconds   D. Approximately 2000 seconds

The answer is C. See answers 4BE-2B.18 through 4BE-2B.21. In the fourth Time Constant, the voltage will fall to 36.8% of 39.9 volts.

$$39.9 \text{ volts X } .368 = 14.68 \text{ volts.}$$

It will, therefore, take approximately four Time Constants or 4 x 450 seconds (1800 seconds) to fall to 14.8 volts.

4BE-3.1 What is a <u>Smith Chart</u>?
A. A graph for calculating impedance along transmission lines
B. A graph for calculating great circle bearings
C. A graph for calculating antenna height
D. A graph for calculating radiation patterns

The answer is A. A Smith Chart is a graphical chart used to analyze and solve problems involving antennas, transmission lines and matching stubs. It is useful in determining the following: (1) the impedance of an antenna at or off resonance, (2) the input impedance

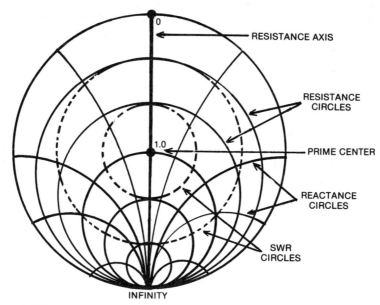

Figure 4BE-3.1. A simplified Smith Chart. The vertical line represents the resistance axis. The circles that are all tangent to the bottom of the chart are the resistance circles. The segments that are all starting at a point on the bottom and curving to the right and left represent reactance circles. The dashed line circles centering on the Prime Center are the SWR circles.

of a transmission line when not terminated at its characteristic impedance, (3) SWR and line losses, and (4) required lengths of matching stubs to properly match a transmission line to an antenna.

The Smith Chart consists of a circular graph of curved lines and concentric circles which represent such factors as resistance, inductive and capacitive reactance, standing wave ratio and wavelength. When two or more factors are known and plotted, the other related factors can be read from the chart. See Figure 4BE-3.1 for a diagram and further explanation of a Smith Chart.

4BE-3.2 What type of coordinate system is used in a <u>Smith Chart</u>?
A. Voltage and current circles
B. Resistance and reactance circles
C. Voltage and current lines
D. Resistance and reactance lines
    The answer is B. See answer 4BE-3.1.

4BE-3.3 What type of calculations can be performed using a <u>Smith Chart</u>?
A. Beam headings and radiation patterns
B. Satellite azimuth and elevation bearings
C. Impedance and SWR values in transmission lines
D. Circuit gain calculations
    The answer is C. See answer 4BE-3.1.

4BE-3.4 What are the two families of circles that make up a <u>Smith Chart</u>?
A. Resistance and voltage   B. Reactance and voltage
C. Resistance and reactance   D. Voltage and impedance
    The answer is C. See answer 4BE-3.1.

4BE-3.5 What is the only straight line on a blank <u>Smith Chart</u>?
A. The reactance axis   B. The resistance axis
C. The voltage axis   D. The current axis
    The answer is B. The resistance axis is the only straight line on the chart. It forms the diameter of the chart and is assigned values of zero at one end, up to infinity at the other end. The center of the resistance axis, which is the center of the chart, is assigned a value of 1.0. The exact center of the chart is called the PRIME CENTER.

4BE-3.6 What is the process of <u>normalizing</u> with regard to a Smith Chart?
A. Reassigning resistance values with regard to the reactance axis
B. Reassigning reactance values with regard to the resistance axis
C. Reassigning resistance values with regard to the prime center
D. Reassigning prime center with regard to the reactance axis
    The answer is C. It was pointed out in answer 4BE-3.5 that the resistance axis has values of from zero to infinity, with 1.0 being at the prime center. If we want to use the chart with a 50 ohm line, we can assign a value of 50 ohms to the prime center which had an "original" or "normal" value of 1.0. The original 2.0 would then be 100 ohms, the original or normal 0.5 would become 25 ohms, etc. If we wish

to convert the 50 ohms, or 100 ohms or 25 ohms back to the normal values of 1.0, 2.0 or 0.5, etc., we simply divide the resistance value by 50. This process is called NORMALIZING.

4BE-3.7 What are the curved lines on a <u>Smith Chart</u>?
A. Portions of current circles   B. Portions of voltage circles
C. Portions of resistance circles   D. Portions of reactance circles
   The answer is D. By "curved lines" we mean the lines starting at the bottom of the Smith Chart and curving up to the right and left. This is in contrast to the complete resistance circles. See the Smith Chart in Figure 4BE-3.1.

4BE-3.8 What is the third family of circles, which are added to a Smith Chart during the process of solving problems?
A. Coaxial length circles   B. Antenna length circles
C. Standing wave ratio circles   D. Radiation pattern circles
   The answer is C. SWR circles are not on the basic Smith Chart. They are added when solving problems involving SWR. The SWR circles are centered on the Prime Center of the chart, and each point on a particular circle represents the same SWR values. See the Smith Chart in Figure 4BE-3.1.

4BE-3.9 How are the <u>wavelength scales</u> on a Smith Chart calibrated?
A. In portions of transmission line electrical frequency
B. In portions of transmission line electrical wavelength
C. In portions of antenna electrical wavelength
D. In portions of antenna electrical frequency
   The answer is B. Wavelength scales are plotted around the perimeter of the chart, and are calibrated in terms of portions of the electrical wavelength along a transmission line.

4BE-4.1 What is the impedance of a network comprised of a 0.1-microhenry inductor in series with a 20-ohm resistor, at 30 MHz? (Specify your answer in rectangular coordinates.)
A. 20 + j19
B. 20 - j19
C. 19 + j20
D. 19 - j20

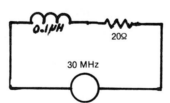

Fig. 4BE-4.1
   The answer is A. In order to find the impedance, we must find the inductive reactance.

$$X_L = 2\pi fL = 6.28 \times 30,000,000 \times .0000001 = 18.84 \text{ ohms.}$$

We then substitute X_L and R in the impedance formula to find the impedance.

$$Z = \sqrt{R^2 + X_L^2} = \sqrt{20^2 + 18.84^2} = \sqrt{754.9} = 27.47 \text{ ohms.}$$

The impedance is therefore 27.47 ohms.

The impedance value of 27.47 ohms does not tell us whether the reactance component is inductive or capacitive. There is another way of describing the impedance of a network containing resistance and reactance that gives us some additional information. It is called the "rectangular form". In this method, the basic impedance formula is:

$$Z = R \pm jX$$ where: Z is the impedance, R is the resistance, X is the reactance and j is the square root of $-1 (\sqrt{-1})$.

j is also referred to as the j operator. The j operator in front of the X tells us that the X is 90 degrees out of phase with R. It means that R and X cannot simply be added to arrive at the total numerical impedance. We use a plus (+) in front of the j to indicate that the reactance is inductive, and a minus (-) in front of the j to indicate that the reactance is capacitive.

The answer is 20 + j19 because R = 20 ohms, and X, the inductive reactance, is 19 (18.84) ohms. The j tells us that the 19 ohms is a reactance and the + tells us that the reactance is inductive.

4BE-4.2 What is the impedance of a network comprised of a 0.1-microhenry inductor in series with a 30-ohm resistor, at 5 MHz? (Specify your answer in rectangular coordinates.)
A. 30 – j3   B. 30 + j3   C. 3 + j30   D. 3 – j30

The answer is B. Using the same reasoning as in answer 4BE-4.1, we first find the inductive reactance.

$$X_L = 6.28 \times 5,000,000 \times .0000001 = 3.14 \text{ ohms}$$

Using the second method of answer 4BE-4.1 to describe impedance, we have:
$$Z = R + jX = 30 + j3$$

We use the + in front of the j because the reactance is inductive.

4BE-4.3 What is the impedance of a network comprised of a 10-microhenry inductor in series with a 40-ohm resistor, at 500 MHz? (Specify your answer in rectangular coordinates.)
A. 40 + j31400   B. 40 – j31400   C. 31400 + j40   D. 31400 – j40

The answer is A. Using the same reasoning as in answer 4BE-4.1, we first find the inductive reactance.

$$X_L = 6.28 \times 5,000,000 \times .00001 = 31,400 \text{ ohms}$$

Using the second method of answer 4BE-4.1 to describe impedance, we have:
$$Z = R + jX = 40 + j31,400$$

We use the + in front of the j because the reactance is inductive.

4BE-4.4 What is the impedance of a network comprised of a 100-

picofarad capacitor in parallel with a 4000-ohm resistor, at 500 kHz?
(Specify your answer in polar coordinates.)
A. 2490 ohms, / 51.5 degrees   B. 4000 ohms, / 38.5 degrees
C. 5112 ohms, / -38.5 degrees   D. 2490 ohms, / -51.5 degrees

The answer is D. Unlike the previous questions, this question involves a parallel circuit and requires that the answer be in "polar coordinates, not "rectangular coordinates". In "polar coordinates" we list the numerical impedance and the phase angle between the voltage and current.

We first find the capacitive reactance of the capacitor using the following formula:

$$X_c = \frac{10^6}{2\pi\, fc}$$

where: X_c is the capacitive reactance in ohms
f is the frequency in megaHertz
c is the capacity in picofarads
π is 3.14

$$X_c = \frac{10^6}{6.28 \times .5 \times 100} = \frac{1,000,000}{314} = 3184.7 \text{ ohms}$$

In order to find the total numerical impedance of a parallel circuit containing a reactance, X and a resistance, R, we use the following formula:

$$Z = \frac{RX}{\sqrt{R^2 + X^2}}$$

Substituting the values in this problem, we get

$$Z = \frac{4000 \times 3185}{\sqrt{4000^2 + 3185^2}} = \frac{12740000}{\sqrt{26144225}} = 2491.6 \text{ ohms}$$

Though there are many complex methods of finding the phase angle, a simple one for a parallel circuit of this type is:

$$\text{Tangent } \theta = \frac{R}{X}$$

where: θ is the phase angle
R is the resistance
X is the reactance

$$\text{Tan } \theta = \frac{4000}{3185} = 1.256$$

The trigonometry table tells us that 1.256 is the tangent of 51.5 degrees. Thus, in the polar coordinate form, the impedance is 2490 ohms /-51.5 degress. The minus (-) in front of the angle tells us that the reactance is capacitive.

4BE-4.5 What is the impedance of a network comprised of a 0.001-microfarad capacitor in series with a 400-ohm resistor, at 500 kHz? (Specify your answer in rectangular coordinates.)
A. 400 - j318   B. 318 - j400   C. 400 + j318   D. 318 + j400

The answer is A. This problem is solved in a manner similar to question 4BE-4.1. First we find the capacitive reactance.

$$X_c = \frac{10^6}{2 \pi fC} = \frac{10^6}{6.28 \times .5 \times 1000} = \frac{1,000,000}{3140} = 318.4$$

Note that the frequency and capacity were changed to MHz and pf respectively.

Using the second method of answer 4BE-4.1 to describe impedance, we have:

$$Z = R \pm jX = 400 - j318$$

We use the minus (-) sign because the reactance is capacitive.

4BE-5.1 What is the impedance of a network comprised of a 100-ohm-reactance inductor in series with a 100-ohm resistor? (Specify your answer in polar coordinates.)

A. 121 ohms, $/\underline{35 \text{ degrees}}$
B. 141 ohms, $/\underline{45 \text{ degrees}}$
C. 161 ohms, $/\underline{55 \text{ degrees}}$
D. 181 ohms, $/\underline{65 \text{ degrees}}$

Fig. 4BE-5.1

The answer is B. See Figure 4BE-5.1. We use the series impedance formula to find the circuit impedance.

$$Z = \sqrt{R^2 + X^2} = \sqrt{100^2 + 100^2} = \sqrt{20,000} = 141.42 \text{ ohms}$$

We use the following formula to find the phase angle in a series circuit.

$$\text{Tangent } \theta = \frac{X}{R}$$
where: θ (theta) is the phase angle
X is the reactance
R is the resistance

$$\tan \theta = \frac{100}{100} = 1$$

The trigonometry table tells us that the tangent of 45 degrees is 1. Thus, the phase angle is equal to 45 degrees. The impedance (in polar coordinates) is 141 ohms $/\underline{45 \text{ degrees}}$. The absence of a "-" in front of the angle indicates that the reactance is inductive.

4BE-5.2 What is the impedance of a network comprised of a 100-ohm-reactance inductor, a 100-ohm-reactance capacitor, and a 100-ohm resistor all connected in series? (Specify your answer in polar coordinates.)

A. 100 ohms, $/\underline{90 \text{ degrees}}$　　　B. 10 ohms, $/\underline{0 \text{ degrees}}$
C. 100 ohms, $/\underline{0 \text{ degrees}}$　　　D. 10 ohms, $/\underline{100 \text{ degrees}}$

The answer is C. Since the inductive reactance is equal and opposite to the capacitive reactance, they cancel each other out, leaving zero reactance. The only resistance left in the circuit is the actual 100 ohm resistor. Therefore, the impedance is 100 ohms.

A simple way of understanding the 0 degree phase angle is to realize that if there is no reactance, the voltage and current are IN PHASE and therefore, there is NO phase angle between them.

4BE-5.3 What is the impedance of a network comprised of a 400-

ohm-reactance capacitor in series with a 300-ohm resistor? (Specify
your answer in polar coordinates.)
A. 240 ohms, /36.9 degrees   B. 240 ohms, / -36.9 degrees
C. 500 ohms, / 53.1 degrees   D. 500 ohms, / -53.1 degrees
   The answer is D.We use the same method to solve this problem
as in answer 4BE-5.1.

$$Z = \sqrt{300^2 + 400^2} = \sqrt{250000} = 500 \text{ ohms}$$

$$\text{Tan } \theta = \frac{400}{300} = 1.333$$

From the trigonometry table, we find that the tangent of 53 degrees
is equal to 1.33. Therefore, the phase angle, Theta, is -53 degrees,
the minus (-) indicating a capacitive reactance.

**4BE-5.4 What is the impedance of a network comprised of a 300-
ohm-reactance capacitor, a 600-ohm-reactance inductor, and a 400-
ohm resistor, all connected in series? (Specify your answer in polar
coordinates.)**
A. 500 ohms, / 37 degrees   B. 400 ohms, / 27 degrees
C. 300 ohms, / 17 degrees   D. 200 ohms, / 10 degrees
   The answer is A. The impedance formula that we use for a circuit
containing R, L and C is:

$$Z = \sqrt{R^2 + (X_L - X_C)^2}$$

Substituting the given values, we get:
$$Z = \sqrt{400^2 + (600 - 300)^2} = \sqrt{400^2 + (300)^2} = \sqrt{250,000} = 500 \text{ ohms}$$

We use the same formula as in answer 4BE-5.1 to find the phase
angle. However, if we have both inductive and capacitive reactances,
we subtract one from the other to find the resultant reactance. This
is because the reactances are opposite to one another and tend to
cancel each other out. The resultant reactance in this problem is 600 -
300 or 300 ohms.

$$\tan \theta = \frac{X}{R} = \frac{300}{400} = .75$$

From the trigonometry table, we find that the tangent of 37 degrees
is .75. Therefore, the phase angle, Theta, is 37 degrees. Since the
resultant reactance is inductive, we do not put a sign in front of the
37 degrees. This is the same as a plus (+), which is what we use for
an inductive reactance.

**4BE-5.5 What is the impedance of a network comprised of a 400-
ohm-reactance inductor in parallel with a 300-ohm resistor? (Specify
your answer in polar coordinates.)**
A. 240 ohms, / 36.9 degrees   B. 240 ohms, / -36.9 degrees
C. 500 ohms, / 53.1 degrees   D. 500 ohms, / -53.1 degrees

The answer is A. We use a similar method to solve this problem as in answer 4BE-4.4. The numerical impedance of the parallel circuit is:

$$Z = \frac{RX}{\sqrt{R^2 + X^2}} = \frac{300 \times 400}{\sqrt{300^2 + 400^2}} = \frac{120,000}{\sqrt{250,000}} = 240 \text{ ohms}$$

The phase angle for this parallel circuit is:

$$\text{Tan } \theta = \frac{R}{X} = \frac{300}{400} = .75$$

From the trigonometry table we see that the tangent of 36.9 degrees is .75. The impedance in polar form is therefore 240 ohms $\underline{/\ 36.9}$ degrees. The absence of "–" in front of the angle indicates that the reactance is inductive.

4BE-6A.1 What is the impedance of a network comprised of a 1.0-millihenry inductor in series with a 200-ohm resistor, at 30 kHz? (Specify your answer in rectangular coordinates.)
A. 200 – j188   B. 200 + j188   C. 188 + j200   D. 188 – j200
The answer is B. Using the same reasoning as in answer 4BE-4.1, we first find the inductive reactance.

$$X_L = 6.28 \times 30,000 \times .001 = 188.4 \text{ ohms}$$

Using the second method of answer 4BE-4.1 to describe impedance, we have:
$$Z = R + jX = 200 + j188$$

We use the + in front of the j because the reactance is inductive.

4BE-6A.2 What is the impedance of a network comprised of a 10-millihenry inductor in series with a 600-ohm resistor, at 10 kHz? (Specify your answer in rectangular coordinates.)
A. 628 + j600   B. 628 – j600   C. 600 + j628   D. 600 – j628
The answer is C. Using the same reasoning as in answer 4BE-4.1, we first find the inductive reactance.

$$X_L = 6.28 \times 10,000 \times .01 = 628 \text{ ohms}$$

Using the second method of answer 4BE-4.1 to describe impedance, we have:
$$Z = R + jX = 600 + j628$$

We use the + in front of the j because the reactance is inductive.

4BE-6A.3 What is the impedance of a network comprised of a 0.01-microfarad capacitor in parallel with a 300-ohm resistor, at 50 kHz? (Specify your answer in rectangular coordinates.)
A. 150 – j159   B. 150 + j159   C. 159 + j150   D. 159 – j150
The answer is D. We first find the capacitive reactance in the circuit, using the formula of answer 4BE-4.5:

$$X_c = \frac{10^6}{6.28 \times .05 \times 10,000} = \frac{1,000,000}{3140} = 318.47 \text{ ohms}$$

The total numerical impedance is equal to:

$$Z = \frac{300 \times 318}{\sqrt{300^2 + 318^2}} = \frac{95,400}{\sqrt{191124}} = \frac{95,400}{437.17} = 218.22 \text{ ohms}$$

The phase angle in this parallel circuit is equal to:

$$\text{Tan } \theta = \frac{R}{X} = \frac{300}{318.4} = .942$$

From the trigonometry table, we see that the tangent of 43.3 degrees is equal to .942. Thus, the phase angle is 43.3 degrees. The impedance, in polar coordinates, is therefore 218.22 / −43.3 degrees.

The question requires that the answer be in rectangular form. In order to convert polar form to rectangular form, we use the following formulas:

$$\text{Rrect} = \text{Zp} \times \cos \theta = 218.22 \times .728 = 158.86$$

$$\text{Xrect} = -j \times \text{Zp} \times \sin \theta = -j218.22 \times .686 = -j149.69$$

Where Rrect is the equivalent resistance in recantgular form, Zp is the total measured impedance, θ is the phase angle and Xrect is the equivalent reactance in rectangular form. The impedance in rectangular form is therefore 159 − j150.

4BE-6A.4 What is the impedance of a network comprised of a 0.1-microfarad capacitor in series with a 40-ohm resistor, at 50 kHz? (Specify your answer in rectangular coordinates.)
A. 40 + j32   B. 40 − j32   C. 32 − j40   D. 32 + j40

The answer is B. This question is similar to question 4BE-4.5. First we find the capacitive reactance.

$$X_c = \frac{10^6}{6.28 \times .05 \times 100,000} = \frac{1,000,000}{31400} = 31.85 \text{ ohms}$$

From answer 4BE-4.5, we know that the formula for the rectangular form notation is:

$$Z = R \pm jX$$

Therefore, Z = 40 − j32. The minus (−) in front of the j indicates that the reactance is capacitive.

4BE-6A.5 What is the impedance of a network comprised of a 1.0-microfarad capacitor in parallel with a 30-ohm resistor, at 5 MHz? (Specify your answer in rectangular coordinates.)
A. 0.000034 + j.032   B. 0.032 + j.000034
C. 0.000034 − j.032   D. 0.032 − j.000034

The answer is C. We use the same method for solving this problem as in answer 4BE-6A-3.

$$X_c = \frac{10^6}{6.28 \times 5 \times 1,000,000} = \frac{1,000,000}{31,400,000} = .0318 \text{ ohms}$$

The total numerical impedance is equal to:

$$Z = \frac{30 \times .0318}{\sqrt{30^2 + .0318^2}} = \frac{.954}{\sqrt{900.001}} = \frac{.954}{30} = .0318 \text{ ohms}$$

We then find the phase angle in this parallel circuit.

$$\text{Tan } \theta = \frac{R}{X} = \frac{30}{.0318} = 943$$

From the trigonometry table, we see that the tangent of 89.94 degrees is equal to 943. Thus, the impedance, in the polar coordinate form, is approximately .0318 /−89.9 degrees.

We convert the polar form to the rectangular form using the formulas of answer 4BE-6A.3.

$$\text{Rrect} = Zp \times \cos \theta = .0318 \times .00107 = .000034$$

$$\text{Xrect} = -j \times Zp \times \sin \theta = -j \times .0318 \times 1 = -j.032$$

Therefore, $Z = .000034 - J.032$. The minus (−) in front of the j indicates that the reactance is capacitive.

4BE–6B.1 What is the impedance of a network comprised of a 100–ohm–reactance capacitor in series with a 100–ohm resistor? (Specify your answer in polar coordinates.)
A. 121 ohms, / −25 degrees   B. 141 ohms, / −45 degrees
C. 161 ohms, / −65 degrees   D. 191 ohms, / −85 degrees
    The answer is B. This is exactly the same question as 4BE–5.1. However, since the reactance is capacitive instead of inductive, we place a minus (−) in front of the phase angle. See answer 4BE–5.1.

4BE–6B.2 What is the impedance of a network comprised of a 100–ohm–reactance capacitor in parallel with a 100–ohm resistor? (Specify your answer in polar coordinates.)
A. 31 ohms, / −15 degrees   B. 51 ohms, / −25 degrees
C. 71 ohms, / −45 degrees   D. 91 ohms, / −65 degrees
    The answer is C. The question is similar to 4BE–5.5 and we use the same method to solve it. The numerical impedance is:

$$Z = \frac{100 \times 100}{\sqrt{100^2 + 100^2}} = \frac{10,000}{141.42} = 70.71 \text{ ohms}$$

We use the formula of 4BE–5.5 to find the phase angle:

$$\tan \theta = \frac{100}{100} = 1$$

From the trigonometry table, we find that the tangent of 45 degrees is equal to 1. Therefore, the phase angle, Theta, is 45 degrees. The impedance, in polar form, is therefore 71 ohms, /-45 degrees, the minus indicating a capacitive reactance.

4BE–6B.3 What is the impedance of a network comprised of a 300-ohm-reactance inductor in series with a 400-ohm resistor? (Specify your answer in polar coordinates.)
A. 400 ohms, / 27 degrees   B. 500 ohms, / 37 degrees
C. 600 ohms, / 47 degrees   D. 700 ohms, / 57 degrees

The answer is B. We use the same method in solving this problem as in answer 4BE–5.1.

$$Z = \sqrt{400^2 + 300^2} = \sqrt{250000} = 500 \text{ ohms}$$

$$\text{Tan } \theta = \frac{300}{400} = .75$$

From the trigonometry table, we find that the tangent of 37 degrees is equal to .75. Therefore, the phase angle, Theta, is 37 degrees.

4BE–6B.4 What is the impedance of a network comprised of a 100-ohm-reactance inductor in parallel with a 100-ohm resistor? (Specify your answer in polar coordinates.)
A. 71 ohms, / 45 degrees   B. 81 ohms, / 55 degrees
C. 91 ohms, / 65 degrees   D. 100 ohms, / 75 degrees

The answer is A. This problem is similar to 4BE–6B.2, and we use the same method to solve it.

$$Z = \frac{100 \times 100}{\sqrt{100^2 + 100^2}} = \frac{10000}{141.42} = 70.71 \text{ ohms}$$

We use the formula of 4BE–5.5 to find the phase angle:

$$\text{Tan } \theta = \frac{100}{100} = 1$$

From the trigonometry table, we find that the tangent of 45 degrees is 1. Therefore, the phase angle, Theta, is 45 degrees.

4BE–6B.5 What is the impedance of a network comprised of a 300-ohm-reactance capacitor in series with a 400-ohm resistor? (Specify your answer in polar coordinates.)
A. 200 ohms, / -10 degrees   B. 300 ohms, / -17 degrees
C. 400 ohms, / -27 degrees   D. 500 ohms, / -37 degrees

The answer is D. This question is the same as 4BE–6B.3, except that the reactance is capacitive. We therefore have a "–" in front of the phase angle whereas the answer in 4BE–6B.3 does not.

4BF-1A.1 What is an enhancement-mode FET?

A. An FET with a channel that blocks voltage through the gate
B. An FET with a channel that allows a current when the gate voltage is zero
C. An FET without a channel to hinder current through the gate
D. An FET without a channel; no current occurs with zero gate voltage

The answer is D. The enhancement-mode FET, in a sense, acts in an opposite manner to the depletion-mode FET. The enhancement-mode FET does NOT have a permanent P or N channel without a gate voltage. It "forms" a channel only when a voltage is applied to the gate. Hence, when there is no gate voltage, there is no channel and there is no current flow. JFETs cannot be used as enhancement-mode FETs. Only IGFETs (MOSFET) can be used.

4BF-1B.1 What is a depletion-mode FET?

A. An FET that has a channel with no gate voltage applied; a current flows with zero gate voltage
B. An FET that has a channel that blocks current when the gate voltage is zero
C. An FET without a channel; no current flows with zero gate voltage
D. An FET without a channel to hinder current through the gate

The answer is A. A depletion-mode FET has an existing P or N channel without a voltage applied to its gate. Source-to-Drain current will flow with zero voltage on the gate. When we apply a negative voltage to the gate, the conduction channel is reduced and less current flows to the drain.

4BF-1C.1 What is the schematic symbol for an N-channel MOSFET?

The answer is A.

4BF-1C.2 What is the schematic symbol for a P-channel MOSFET?

C.  D.

The answer is B. The only difference between the schematic symbol of the N-channel MOSFET and the schematic symbol of the P-channel MOSFET is the direction of the arrow.

4BF-1C.3 What is the schematic symbol for an N-channel dual-gate MOSFET?

A.   B.

C.   D.

The answer is C. Note that this is similar to the single gate MOSFET of questions 4BF-1C.1 and 4BF-1C.2, except that it has two gates.

4BF-1C.4 What is the schematic symbol for a P-channel dual-gate MOSFET?

A.  B.

C.   D.

The answer is D. The only difference between the schematic symbol of the P-channel dual-gate MOSFET and the schematic symbol of the N-channel dual gate MOSFET is the direction of the arrow.

4BF-1C.5 Why do many MOSFET devices have built-in gate-protective Zener diodes?
A. The gate-protective Zener diode provides a voltage reference to provide the correct amount of reverse-bias gate voltage
B. The gate-protective Zener diode protects the substrate from excessive voltages
C. The gate-protective Zener diode keeps the gate voltage within specifications to prevent the device from overheating
D. The gate-protective Zener diode prevents the gate insulation from being punctured by small static charges or excessive voltages

The answer is D. Most MOSFETs manufactured today have built-in protection against static buildup. The protection consists of Zener diodes connected between the gate and the source.

4BF-1D.1 What do the initials CMOS stand for?

A. Common mode oscillating system
B. Complementary mica-oxide silicon
C. Complementary metal-oxide semiconductor
D. Complementary metal-oxide substrate

The answer is C. MOSFET is also referred to as an Insulated Gate FET. CMOS technology allows both P-channel and N-channel devices to share the same silicon substrate. Both devices are enhancement-mode, and only one device is used at a time.

4BF-1D.2 Why are special precautions necessary in handling FET and CMOS devices?
A. They are susceptible to damage from static charges
B. They have fragile leads that may break off
C. They have micro-welded semiconductor junctions that are susceptible to breakage
D. They are light sensitive

The answer is A. The insulation around the gate is thin and is easily damaged by static charges that may accumulate. Therefore, CMOS devices must be handled carefully. They must be stored with their pins touching a conductive circuit, such as aluminum foil or conductive foam. They should never be stored in non-conductive trays or plastic bags. The pin leads should not be touched with the fingers. Instead, the devices should be handled by their body.

4BF-1E.1 What is the schematic symbol for an N-channel junction FET?

The answer is A. G stands for Gate. D stands for Drain and S stands for Source. Figure 4BF-1E.1 illustrates a cross-section of an N-channel junction FET.

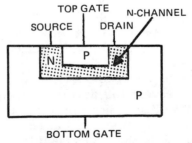

JUNCTION FET (CROSS SECTION)

Fig. 4BF-1E.1. Cross section of a junction FET.

4BF-1E.2 How does the input impedance of a <u>field-effect transistor</u>

compare with that of a bipolar transistor?

A. One cannot compare input impedance without first knowing the supply voltage
B. An FET has low input impedance; a bipolar transistor has high input impedance
C. The input impedance of FETs and bipolar transistors is the same
D. An FET has high input impedance; a bipolar transistor has low input impedance

The answer is D. The FET has a relatively high internal resistance, similar to a vacuum tube, and operates much like a vacuum tube. The source of the FET can be compared to the cathode, the drain to the plate, and the gate to the grid. Because of its high internal resistance, the FET provides less loading and has a better dynamic range than the conventional bipolar transistor. FETs can handle large signals without overloading. Matching and coupling are easier to accomplish with FETs than bipolar units.

4BF-1E.3 What are the three terminals of a field-effect transistor?
A. Gate 1, gate 2, drain   B. Emitter, base, collector
C. Emitter, base 1, base 2   D. Gate, drain, source

The answer is D. See answer 4BF-1E.1.

4BF-1F.1 What is the schematic symbol for a P-channel junction FET?

The answer is B. The only difference between the schematic symbol of the P-channel junction FET and the schematic symbol of the N-channel junction FET is the difference in the direction of the arrow in the Gate.

4BF-1F.2 What are the two basic types of junction field-effect transistors?
A. N-channel and P-channel   B. High power and low power
C. MOSFET and GaAsFET   D. Silicon FET and germanium FET

The answer is A. The supply polarities of an N-channel FET are opposite to those of a P-channel FET. We can recognize the difference between N and P channel FETs in schematic circuits by the reversal of the arrows in their gates.

4BF-2.1 What is an operational amplifier?
A. A high-gain, direct-coupled differential amplifier whose characteristics are determined by components external to the amplifier unit
B. A high-gain, direct-coupled audio amplifier whose characteristics

are determined by components external to the amplifier unit
C. An amplifier used to increase the average output of frequency
modulated amateur signals to the legal limit
D. A program subroutine that calculates the gain of an RF amplifier

The answer is A. An operational amplifier (more commonly called
"OP AMP"), is a high gain, direct-coupled amplifier on a single IC
chip. It is linear and has a high input impedance. Figure 4BF-2.1
shows a typical Op-Amp application.

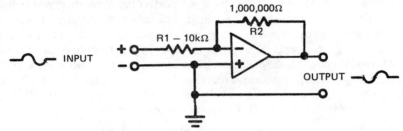

Fig. 4BF-2.1. An op-amp inverting amplifier.

4BF-2.2 What is the schematic symbol for an <u>operational amplifier</u>?

A.

C.

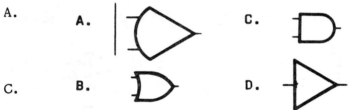

The answer is A.

4BF-2.3 What would be the characteristics of the ideal op-amp?
A. Zero input impedance, infinite output impedance, infinite gain, flat
frequency response
B. Infinite input impedance, zero output impedance, infinite gain, flat
frequency response
C. Zero input impedance, zero output impedance, infinite gain, flat
frequency response
D. Infinite input impedance, infinite output impedance, infinite gain,
flat frequency response

The answer is B. The characteristics given in answer B are ideal
theoretical characteristics, and we try to come as close as possible
to them.

4BF-2.4 What determines the gain of a closed-loop op-amp circuit?
A. The external feedback network
B. The collector-to-base capacitance of the PNP stage
C. The power supply voltage
D. The PNP collector load

The answer is A. The gain of an Op-Amp circuit is determined by
the value of the resistor in the external feedback network. The
formula for finding the gain of the Op-Amp in Figure 4BF-2.1 is:

$$\text{GAIN} = \frac{R1 + R2}{R1}$$

The gain of the Op-Amp of Figure 4BF-2.1 is:

$$\text{GAIN} = \frac{10,000 + 1,000,000}{10,000} = 101$$

4BF-2.5 What is meant by the term op-amp offset voltage?

A. The output voltage of the op-amp minus its input voltage
B. The difference between the output voltage of the op-amp and the input voltage required in the following stage
C. The potential between the amplifier-input terminals of the op-amp in a closed-loop condition
D. The potential between the amplifier-input terminals of the op-amp in an open-loop condition

The answer is C. It is the voltage measured between the two input terminals in the closed loop condition. This voltage should normally be zero; however, an extremely small voltage is usually detectable due to variations in parts inside of the OP AMP.

4BF-2.6 What is the input impedance of a theoretically ideal op-amp?

A. 100 ohms   B. 1000 ohms   C. Very low   D. Very high

The answer is D. See questions 4BF-2.1 through 4BF-2.3.

4BF-2.7 What is the output impedance of a theoretically ideal op-amp?

A. Very low   B. Very high   C. 100 ohms   D. 1000 ohms

The answer is A. See questions 4BF-2.1 through 4BF-2.3.

4BF-3.1 What is a phase-locked loop circuit?

A. An electronic servo loop consisting of a ratio detector, reactance modulator, and voltage-controlled oscillator
B. An electronic circuit also known as a monostable multivibrator
C. An electronic circuit consisting of a precision push-pull amplifier with a differential input
D. An electronic servo loop consisting of a phase detector, a low-pass filter and voltage-controlled oscillator

The answer is D. The overall purpose of a phase-locked loop is to develop a stable signal. It does this by comparing the signal to a reference and correcting the signal accordingly.

A phase-locked loop (PLL) integrated circuit, contains the circuit functions shown in Figure 4BF-3.1. The voltage controlled oscillator (VCO) will change its output frequency in response to a change in its applied voltage. A "sample" of the output frequency of the VCO is fed back into a phase detector circuit. Here it is compared to a reference frequency which is generated by the reference oscillator. The phase detector generates a voltage proportional to the frequency difference between the VCO and the reference signal. This voltage is then applied to a low pass filter to remove undesired voltages from the output of the phase detector. The output of the low pass filter is applied to the VCO, where it changes the VCO frequency to bring it back in line with the reference oscillator. This allows the VCO to "lock" or synchronize its output with the reference signal.

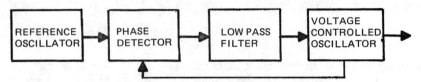

Fig. 4BF-3.1. A phase-locked loop IC.

4BF-3.2 What functions are performed by a phase-locked loop?
A. Wideband AF and RF power amplification
B. Comparison of two digital input signals, digital pulse counter
C. Photovoltaic conversion, optical coupling
D. Frequency synthesis, FM demodulation

The answer is D. See question 4BF-3.1. A number of manufacturers make single chip IC's that contain all the circuits of the PLL. PLL IC's can be used as FM detectors, since the error voltage produced by the phase detector is a duplicate of the audio voltage used to produce the frequency shift of an FM signal. Other applications include frequency synthesis and synchronization. Since no mixing is used in a PLL, the output is relatively "clean" of harmonics and other unwanted emissions.

4BF-3.3 A circuit compares the output from a voltage-controlled oscillator and a frequency standard. The difference between the two frequencies produces an error voltage that changes the voltage-controlled oscillator frequency. What is the name of the circuit?
A. A doubly balanced mixer   B. A phase-locked loop
C. A differential voltage amplifier
D. A variable frequency oscillator

The answer is B. See question 4BF-3.1

4BF-4.1 What do the initials TTL stand for?
A. Resistor-transistor logic   B. Transistor-transistor logic
C. Diode-transistor logic   D. Emitter-coupled logic

The answer is B. Transistor-transistor logic (TTL), is a form of digital logic which makes use of the switching properties of transistor circuits. TTL circuitry is known for the relative high speed with which its digital operation takes place.

4BF-4.2 What is the recommended power supply voltage for TTL series integrated circuits?
A. 12.00 volts   B. 50.00 volts   C. 5.00 volts   D. 13.60 volts

The answer is C. The 5 volts should not be exceeded; otherwise the device may be damaged.

The 7400 series of integrated circuits are the basic TTL IC's. They contain several digital logic circuits, including OR, AND, NOR, NAND and flip-flop circuits in a single chip. The "7400 series" includes many devices having a part number beginning with "74" and including two

other digits. Figure 4BF-4.2 shows the internal construction of the
7400. Note that the 7400 includes four NAND gates on the same chip.
It is widely used since AND, OR and NOR devices can be built from
combinations of NAND gates.

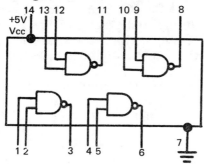

Fig. 4BF-4.2. A QUAD NAND Gate 7400 TTL IC.

**4BF-4.3 What logic state do the inputs of a TTL device assume if
they are left open?**
A. A high logic state
B. A low logic state
C. The device becomes randomized and will not provide consistent
high or low logic states
D. Open inputs on a TTL device are ignored
The answer is A.

**4BF-4.4 What level of input voltage is high in a TTL device operating
with a 5-volt power supply?**
A. 2.0 to 5.5 volts
B. 1.5 to 3.0 volts
C. 1.0 to 1.5 volts
D. -5.0 to -2.0 volts
The answer is A. Any voltage above 2 volts is considered high.

**4BF-4.5 What level of input voltage is low in a TTL device operating
with a 5-volt power supply?**
A. -2.0 to -5.5 volts
B. 2.0 to 5.5 volts
C. -0.6 to 0.8 volts
D. -0.8 to 0.4 volts
The answer is C.

**4BF-4.6 Why do circuits containing TTL devices have several bypass
capacitors per printed circuit board?**
A. To prevent RFI to receivers
B. To keep the switching noise within the circuit, thus eliminating
RFI
C. To filter out switching harmonics
D. To prevent switching transients from appearing on the supply
line
The answer is D. Since TTL devices are used for high speed
switching, it is only natural that switching transients will appear in
the TTL devices. These transients will get into the power supply
lines and cause problems with other devices, unless they are by-
passed with capacitors.

4BF-5.1 What is a CMOS IC?

A. A chip with only P-channel transistors
B. A chip with P-channel and N-channel transistors
C. A chip with only N-channel transistors
D. A chip with only bipolar transistors

The answer is B. "CMOS IC" stands for Complementary Metal-Oxide Semiconductor Integrated Circuit. They contain several digital circuits combined in a single chip. Complementary FETs are used. Thus, the answer to the question indicates P-channel and N-channel transistors.

CMOS ICs consume very little current. This is because they are enhancement-mode devices; only one of the two complementary parts of a circuit section is turned on at a time. CMOS ICs are easily damaged by static electricity. This can be prevented by storing CMOS IC devices with their pins touching conductive material, such as aluminum foil or conductive foam. They should never be stored on non-conductive trays or in plastic bags.

4BF-5.2 What is one major advantage of <u>CMOS</u> over other devices?
A. Small size   B. Low current consumption
C. Low cost   D. Ease of circuit design

The answer is B. The 4000 series CMOS ICs are similar to the 7400 series TTLs in that they have several digital circuits combined in a single chip. However, CMOS ICs can contain more functions in the same size chip than TTL ICs. Vcc for most CMOS devices can range between 3 and 15 volts (in contrast to the narrow requirements of the TTL series). The CMOS IC derives its name from the fact that the circuitry in each chip uses CMOS transistors.

CMOS ICs consume very little current. This is because they are enhancement-mode devices; only one of the two complementary parts of a circuit section is turned on at a time. However, CMOS ICs are generally smaller than TTL devices. CMOS ICs are easily damaged by static electricity. See answer 4BF-5.1 for measures to prevent damage due to static.

4BF-5.3 Why do <u>CMOS</u> digital integrated circuits have high immunity to noise on the <u>input</u> signal or power supply?
A. Larger bypass capacitors are used in CMOS circuit design
B. The input switching threshold is about two times the power supply voltage
C. The input switching threshold is about one-half the power supply voltage
D. Input signals are stronger

The answer is C.

4BF-6.1 What is the name for a vacuum tube that is commonly found in television cameras used for amateur television?
A. A traveling-wave tube   B. A klystron tube
C. A vidicon tube   D. A cathode-ray tube

The answer is C. A vidicon is a tube that is used in a TV camera to transform the visual images to be televised into varying electrical currents. Figure 4BF-6.1 shows its internal construction.

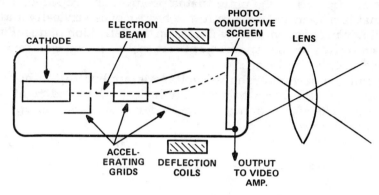

Fig. 4BF–6.1 A vidicon TV Camera Tube.

The images are focused through a lens onto the photo-conductive screen of the vidicon tube. An electron beam is formed at the cathode of the electron gun. The beam is shaped, focused, and aimed at the screen by three accelerating grids. Deflection coils sweep the beam from left to right and, less rapidly, from top to bottom, much like a TV receiver's picture tube. As the beam strikes the photo-sensitive screen, it charges the bombarded area to a certain potential. Then, as soon as the beam moves on, a small current flows from the screen material out to the video amplifier in the external circuit. The amount of current flow at any instant depends on the brightness of the image at that specific point on the screen.

4BF–6.2 How is the electron beam deflected in a vidicon?
A. By varying the beam voltage
B. By varying the bias voltage on the beam forming grids inside the tube
C. By varying the beam current
D. By varying electromagnetic fields
The answer is D. See answer 4BF–6.1.

4BF–6.3 What type of CRT deflection is better when high-frequency waves are to be displayed on the screen?
A. Electromagnetic   B. Tubular
C. Radar   D. Electrostatic
The answer is D. Since electrostatic deflection is better than magnetic deflection for the display of high frequency signals, it is used extensively for oscillograph purposes.

The purpose of a cathode-ray tube (CRT) is to produce a visible display of an electrical waveform. Figure 4BF–6.3 shows the elements in a CRT. The cathode at the left emits electrons, much like any vacuum tube. The electrons are drawn through a small opening in the control grid which controls the amount of electrons in the beam. Next, the electrons pass through two positive anodes. These anodes propel the electrons toward the screen and focus them into a small beam. Next, the beam passes between the vertical and horizontal deflection plates. By placing varying potentials on these plates, the beam at any instant is attracted to the plate that is positive, and repelled from the plate that is negative. As a result, the beam moves vertically and horizontally, in proportion to the voltages on the deflection plates. Finally, the beam strikes the phosphorescent material on the screen, producing a visible line as the beam moves.

Normally, a sawtooth-shaped voltage is applied to the horizontal plates to cause the beam to move rhythmically left and right across the screen. The signal or waveform under observation is applied to the vertical plates.

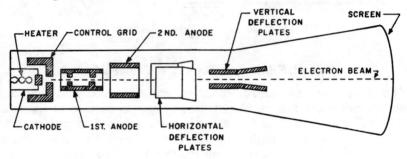

Fig. 4BF–6.3. A cathode-ray tube.

SUBELEMENT 4BG
PRACTICAL CIRCUITS
(4 questions)

4BG-1A.1 What is a flip-flop circuit?
A. A binary sequential logic element with one stable state
B. A binary sequential logic element with eight stable states
C. A binary sequential logic element with four stable states
D. A binary sequential logic element with two stable states

The answer is D. A flip-flop circuit is a bistable multivibrator. A multivibrator is a simple, two-stage RC coupled amplifier, with the output of the second stage fed back to the input of the first stage. Bistable means that there are two stable operating states. It remains in one of its two states indefinitely until an external "trigger" signal is applied. It then switches to the other state and remains there until another "trigger" signal is applied. A basic bistable multivibrator is shown in Figure 4BG-1A.1.

Fig. 4BG-1A.1. A bistable multivibrator.

4BG-1A.2 How many bits of information can be stored in a single flip-flop circuit?
A. 1   B. 2   C. 3   D. 4

The answer is A. Flip-flop circuits are used in memory devices. A single flip-flop circuit can store one bit of memory.

4BG-1A.3 What is a bistable multivibrator circuit?
A. An "AND" gate   B. An "OR" gate
C. A flip-flop   D. A clock

The answer is C. See answer 4BG-1A.1.

4BG-1A.4 How many output changes are obtained for every two trigger pulses applied to the input of a bistable T flip-flop circuit?
A. No output level changes   B. One output level change
C. Two output level changes   D. Four output level changes

The answer is C. Each trigger pulse causes a level output change. Therefore, two trigger pulses will cause two output level changes. See answer 4BG-1A.1.

4BG-1A.5 The frequency of an ac signal can be divided electronically by what type of digital circuit?
A. A free-running multivibrator
B. An OR gate
C. A bistable multivibrator
D. An astable multivibrator

The answer is C. A digital frequency divider circuit makes use of the ability of flip-flop circuits to "divide" inputs. In Figure 4BG-1A.5, a crystal marker circuit is used to illustrate this ability. In a typical application, a 100 kHz crystal controls an oscillator's frequency. The flip-flop circuits are made to produce output on every other pulse of input. Thus, if a 50 kHz output is desired, the 100 kHz signal is applied to the input of a flip-flop. The output at Q of the flip-flop will have a frequency of 50 kHz. The 50 kHz output at Q can be used as the input to a second flip-flop, and the output at Q of the second flip-flop will have a frequency of 25 kHz. Division by two can continue in this manner. The oscillator used in a crystal marker circuit is designed to have a rich harmonic content, so it can be used as a reference at frequencies well into the MegaHertz range.

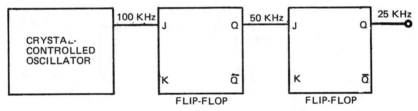

Fig. 4BG-1A.5. A digital frequency divider circuit.

4BG-1A.6 What type of digital IC is also known as a <u>latch</u>?
A. A decade counter   B. An OR gate
C. A flip-flop   D. An op-amp

The answer is C. A "flip-flop" circuit is known as a "latch" because it holds or "latches" onto a certain voltage condition or state. It holds this state until an external trigger voltage causes it to "flip" over to a second state. See answer 4BG-1A.1.

4BG-1A.7 How many <u>flip-flops</u> are required to divide a signal frequency by 4?
A. 1   B. 2   C. 4   D. 8

The answer is B. The first flip-flop divides the signal by 2 and the second flip-flop divides it by another 2, giving an overall division by 4. See answers 4BG-1A.5 and 4BG-2A.2.

4BG-1B.1 What is an <u>astable multivibrator</u>?
A. A circuit that alternates between two stable states
B. A circuit that alternates between a stable state and an unstable state
C. A circuit set to block either a 0 pulse or a 1 pulse and pass the

other

D. A circuit that alternates between two unstable states

The answer is D. An astable multivibrator has no stable states; it continually switches back and forth between two states (off and on) or 0 and 1. Figure 4BG-1B.1 illustrates a basic astable multivibrator. Astable multivibrators are used in timing circuits.

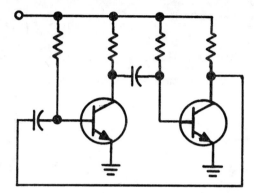

Fig. 4BG-1B.1. An astable multivibrator.

4BG-1B.2 What is a monostable multivibrator?
A. A circuit that can be switched momentarily to the opposite binary state and then returns after a set time to its original state
B. A "clock" circuit that produces a continuous square wave oscillating between 1 and 0
C. A circuit designed to store one bit of data in either the 0 or the 1 configuration
D. A circuit that maintains a constant output voltage, regardless of variations in the input voltage

The answer is A. Monostable oscillators are sometimes referred to as "one shot" multivibrators and are used in radar circuits.

4BG-1C.1 What is an AND gate?
A. A circuit that produces a logic "1" at its output only if all inputs are logic "1"
B. A circuit that produces a logic "0" at its output only if all inputs are logic "1"
C. A circuit that produces a logic "1" at its output if only one input is a logic "1"
D. A circuit that produces a logic "1" at its output if all inputs are logic "0"

The answer is A. Digital logic circuits are those that deal with discrete events falling into two categories: OFF (0 or low) and ON (1 or high). There are no intermediate values between these two events. This is in marked contrast to the infinite number of values that a voltage (for example) could assume in an analog circuit. The term "logic" implies a certain rational behavior for digital circuits, and this happens to be the case.

The AND circuit has two or more inputs and a single output. See

Figure 4BG–1C.1A. The output will be "on" (1) only when all of its inputs are also "on". If any of the inputs are "off" (0), the output will be 0. This is shown in the table of Figure 4BG–1C.1B.

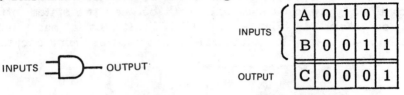

INPUTS	A	0	1	0	1
	B	0	0	1	1
OUTPUT	C	0	0	0	1

INPUTS ─⊐D─ OUTPUT

Fig. 4BG–1C.1A. An AND Gate.   **Fig. 4BG–1C.1B An AND Table**

4BG–1C.2 What is the schematic symbol for an <u>AND gate</u>?

A.

B.

C.

D.

The answer is A. See answer 4BG–1C.1.

4BG–1C.3 What is a <u>NAND gate</u>?
A. A circuit that produces a logic "0" at its output only when all inputs are logic "0"
B. A circuit that produces a logic "1" at its output only when all inputs are logic "1"
C. A circuit that produces a logic "0" at its output if some but not all of its inputs are logic "1"
D. A circuit that produces a logic "0" at its output only when all inputs are logic "1"
The answer is D. The NAND circuit can be thought of as the reverse of the AND circuit. It also has one or more inputs and a single output. If all the inputs are 1, the output is 0. Otherwise, the output is 1. This is shown in the table of Figure 4BG–1C.3B.

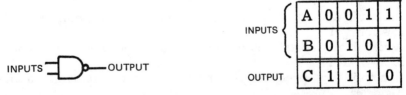

INPUTS	A	0	0	1	1
	B	0	1	0	1
OUTPUT	C	1	1	1	0

INPUTS ─⊐D○─ OUTPUT

Fig. 4BG–1C.3A. A NAND Gate.   **Fig. 4BG–1C.3B. A NAND Table.**

4BG–1C.4 What is the schematic symbol for a <u>NAND gate</u>?

A.   B.

C. 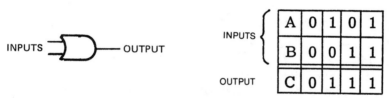   D.

The answer is B. See answer 4BG–1C.3.

4BG–1C.5 What is an OR gate?
A. A circuit that produces a logic "1" at its output if any input is logic "1"
B. A circuit that produces a logic "0" at its output if any input is logic "1"
C. A circuit that produces a logic "0" at its output if all inputs are logic "1"
D. A circuit that produces a logic "1" at its output if all inputs are logic "0"

The answer is A. The OR circuit, like the AND circuit, has two or more inputs and a single output. See Figure 4BG–1C.5A. The output of the OR circuit will be 1 if any or all of the inputs are 1. Its output is 0 only if all of the inputs are 0. This is shown in the table of Figure 4BG–1C.5B.

INPUTS ▭ OUTPUT

INPUTS {

A	0	1	0	1
B	0	0	1	1

OUTPUT

C	0	1	1	1

Fig. 4BG–1C.5A. An OR Gate   Fig. 4BG–1C.5B. An OR Table.

4BG–1C.6 What is the schematic symbol for an OR gate?

A.    B.

C.   D.

The answer is D. See answer 4BG–1C.5.

4BG–1C.7 What is a NOR gate?
A. A circuit that produces a logic "0" at its output only if all inputs are logic "0"
B. A circuit that produces a logic "1" at its output only if all inputs are logic "1"
C. A circuit that produces a logic "0" at its output if any or all inputs are logic "1"
D. A circuit that produces a logic "1" at its output if some but not all of its inputs are logic "1"

The answer is C. The NOR circuit may be thought of as the reverse of the OR circuit. It, too, has a single output and two or more inputs. If both inputs are 0, the output of the NOR circuit will be 1. But if any or all of the inputs are 1, then the output will be 0. This is shown in the table of Figure 4BG-1C.7B.

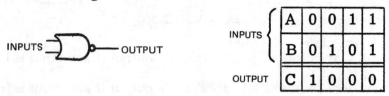

	A	0	0	1	1
INPUTS	B	0	1	0	1
OUTPUT	C	1	0	0	0

Fig. 4BG.1C.7A. A NOR Gate.   Fig. 4BG-1C.7B. A NOR Table.

4BG-1C.8 What is the schematic symbol for a NOR gate?

A.   B.

C.   D.

The answer is D. See answer 4BG-1C.7.

4BG-1C.9 What is a NOT gate?
A. A circuit that produces a logic "O" at its output when the input is logic "1" and vice versa
B. A circuit that does not allow data transmission when its input is high
C. A circuit that allows data transmission only when its input is high
D. A circuit that produces a logic "1" at its output when the input is logic "1" and vice versa

The answer is A. In the NOT gate, the output is the opposite of the input. If the input is "1", the output is "0". If the input is "0", the output is "1". This is shown in Figure 4BG-1C.9B.

INPUT	A	0	1
OUTPUT	C	1	0

Fig. 4BG-1C.9A. A NOT Gate.   Fig. 4BG-1C.9B. A NOT Table.

4BG-1C.10 What is the schematic symbol for a NOT gate?

A.   B.

C.   D.

The answer is A. See answer 4BG-1C.9.

4BG-1D.1 What is a truth table?

A. A table of logic symbols that indicate the high logic states of an op-amp
B. A diagram showing logic states when the digital device's output is true
C. A list of input combinations and their corresponding outputs that characterizes a digital device's function
D. A table of logic symbols that indicates the low logic states of an op-amp

The answer is C. Truth tables give the outputs for the various inputs, and are shown in Figures 4BG-1C.1B through 4BG-1C.9B.

4BG-1D.2 In a positive-logic circuit, what level is used to represent a logic 1?

A. A low level   B. A positive-transition level
C. A negative-transition level   D. A high level

The answer is D. See question 4BG-1C.1. Digital logic circuits also have polarity. In the discussion of question 4BG-1C.1, we assumed that the "high" state was represented by "1" and the "low" state by "0". This is true only if the logic system has positive polarity. When the logic system has negative polarity, the high level is represented by "0" and the low level is represented by "1".

4BG-1D.3 In a positive-logic circuit, what level is used to represent a logic 0?

A. A low level   B. A positive-transition level
C. A negative-transition level   D. A high level

The answer is A. See question 4BG-1D.2.

4BG-1D.4 In a negative-logic circuit, what level is used to represent a logic 1?

A. A low level   B. A positive-transition level
C. A negative-transition level   D. A high level

The answer is A. See question 4BG-1D.2.

4BG-1D.5 In a negative-logic circuit, what level is used to represent a logic 0?

A. A low level   B. A positive-transition level
C. A negative-transition level   D. A high level

The answer is D. See question 4BG-1D.2.

4BG-2A.1 What is a crystal-controlled marker generator?

A. A low-stability oscillator that "sweeps" through a band of frequencies
B. An oscillator often used in aircraft to determine the craft's location relative to the inner and outer markers at airports
C. A high-stability oscillator whose output frequency and amplitude can be varied over a wide range
D. A high-stability oscillator that generates a series of reference signals at known frequency intervals

The answer is D. A frequency marker generator is a highly stable, accurate generator that can be adjusted to put out one or more specific frequency signals. A popular frequency marker uses a 100 kHz crystal in an oscillator circuit that is rich in harmonic output. A considerable number of signals are produced that are 100 kHz apart.

A frequency marker generator can be used in conjunction with a sweep generator to pinpoint specific frequencies in a response curve. It can also be used to check the dial accuracy of a receiver.

4BG-2A.2 What additional circuitry is required in a 100-kHz crystal-controlled marker generator to provide markers at 50 and 25 kHz?

A. An emitter-follower   B. Two frequency multipliers
C. Two flip-flops   D. A voltage divider

The answer is C. One flip-flop will convert the 100 kHz signal to 50 kHz. A second flip-flop will convert the 50 kHz into 25 kHz. See answer 4BG-1A.5.

4BG-2B.1 What is the purpose of a prescaler circuit?

A. It converts the output of a JK flip-flop to that of an RS flip-flop
B. It multiplies an HF signal so a low-frequency counter can display the operating frequency
C. It prevents oscillation in a low frequency counter circuit
D. It divides an HF signal so a low-frequency counter can display the operating frequency

The answer is D. A prescaler consists of frequency dividing circuits that take a high frequency signal and divide it into a frequency low enough for the low frequency counter to read.

4BG-2B.2 What does the accuracy of a frequency counter depend on?

A. The internal crystal reference
B. A voltage-regulated power supply with an unvarying output
C. Accuracy of the ac input frequency to the power supply
D. Proper balancing of the power-supply diodes

The answer is A. Counters make use of the ability of digital circuits to "count". In the block diagram of Figure 4BG-2B.2, the signal to be measured is passed through the circuit preceding the counter IC. This converts the signal into a string of pulses that can be used by a digital circuit. The pulses then are applied to a counter IC. The counter IC needs some standard to compare the input pulses to. This is provided by the time base generator. The time base generator determines the amount of time it takes the pulses representing a certain

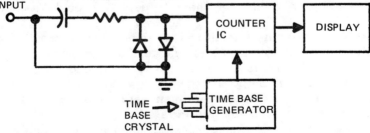

Fig. 4BG-2B.2. A digitial frequency counter circuit.

frequency, usually one MegaHertz, to pass by a counting point. The time base generator's frequency is crystal-controlled. In the counter IC, the time base generator's signal is compared to the input signal, and the input signal pulses are counted in comparison to the time base generator signal. When all the input pulses are counted, the input frequency is shown on an output display.

4BG-2B.3 How many states does a decade counter digital IC have?
A. 6 B. 10 C. 15 D. 20

The answer is B. A decade counter digital IC contains a number of flip-flop circuits, connected in such a way that ten input pulses result in one output pulse. In this manner, a decade counter divides the frequency of a signal by ten. Two decade counters will divide a frequency by one hundred, three decade counters will divide a frequency by one thousand, etc.

4BG-2B.4 What is the function of a decade counter digital IC?
A. Decode a decimal number for display on a seven-segment LED display
B. Produce one output pulse for every ten input pulses
C. Produce ten output pulses for every input pulse
D. Add two decimal numbers

The answer is B. See answer 4BG-2B.3.

4BG-3A.1 What are the advantages of using an op-amp instead of LC elements in an audio filter?
A. Op-amps are more rugged and can withstand more abuse than can LC elements
B. Op-amps are fixed at one frequency
C. Op-amps are available in more styles and types than are LC elements
D. Op-amps exhibit gain rather than insertion loss

The answer is D. Operational amplifiers (OP AMPS) can be used in audio active filters because they have high gain at audio frequencies, and their bandwidths can be controlled by using various values of resistance and capacitance in negative feedback arrangements. Figure 4BG-3A.1 shows the type of negative feedback OP AMP circuit that is used in audio active filters.

On the other hand, LC filters are fixed frequency devices and have a narrow bandwidth. They have an insertion loss and are bulky, compared to modern OP AMP filters.

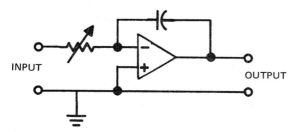

Fig. 4BG-3A.1. A negative feedback OP AMP circuit.

4BG-3A.2 What determines the gain and frequency characteristics of an op-amp RC active filter?
A. Values of capacitances and resistances built into the op- amp
B. Values of capacitances and resistances external to the op- amp
C. Voltage and frequency of dc input to the op-amp power supply
D. Regulated dc voltage output from the op-amp power supply

The answer is B. The gain of an OP AMP is determined by the values of the resistors external to the OP AMP. See the formula for the gain of an OP AMP in answer 4BG-5C.1. The bandwidth and center frequency of the filter can be controlled by the values of the resistors and capacitors.

4BG-3A.3 What are the principle uses of an op-amp RC active filter in amateur circuitry?
A. Op-amp circuits are used as high-pass filters to block RFI at the input to receivers
B. Op-amp circuits are used as low-pass filters between transmitters and transmission lines
C. Op-amp circuits are used as filters for smoothing power- supply output
D. Op-amp circuits are used as audio filters for receivers

The answer is D. The filter is used in the audio section of the receiver to restrict the audio bandwidth of the receiver to the signal being received. In this way, the noise and interference outside of the desired audio band can be greatly reduced.

4BG-3B.1 What type of capacitors should be used in an op-amp RC active filter circuit?
A. Electrolytic   B. Disc ceramic   C. Polystyrene   D. Paper dielectric

The answer is C. Polystyrene capacitors are used because they have a high Q and they are temperature stable.

4BG-3B.2 How can unwanted ringing and audio instability be prevented in a multisection op-amp RC audio filter circuit?
A. Restrict both gain and Q
B. Restrict gain, but increase Q
C. Restrict Q, but increase gain
D. Increase both gain and Q

The answer is A. When the gain and/or Q of a multisection audio active filter becomes too high, a strong "ringing" sound develops. This can be eliminated by reducing the gain and/or Q. If the filter is being built by an amateur, the number of sections should be reduced.

4BG-3B.3 Where should an op-amp RC active audio filter be placed in an amateur receiver?
A. In the IF strip, immediately before the detector
B. In the audio circuitry immediately before the speaker or phone jack
C. Between the balanced modulator and frequency multiplier
D. In the low-level audio stages

The answer is D. Generally speaking, an audio active filter should be inserted in a low level audio stage.

4BG-3B.4 What parameter must be selected when designing an audio filter using an op-amp?
A. Bandpass characteristics   B. Desired current gain
C. Temperature coefficient   D. Output-offset overshoot

The answer is A. Operational amplifiers (OP AMPS) can be used in audio active filters because they have high gain at audio frequencies, and their bandwidths can be controlled by using various values of resistance and capacitance in negative feedback arrangements. Operational amplifiers can be used as high-pass, low-pass and band-pass filters. It is therefore important, when designing an audio active filter, to know the bandpass frequency and the desired voltage gain.

Figure 4BG-3A.1 shows the type of negative feedback OP AMP circuit that is used in audio active filters.

4BG-4A.1 What two factors determine the <u>sensitivity</u> of a receiver?
A. Dynamic range and third-order intercept
B. Cost and availability
C. Intermodulation distortion and dynamic range
D. Bandwidth and noise figure

The answer is D. See answer 4BG-4D.1.

4BG-4A.2 What is the limiting condition for <u>sensitivity</u> in a communications receiver?
A. The noise floor of the receiver
B. The power-supply output ripple
C. The two-tone intermodulation distortion
D. The input impedance to the detector

The answer is A. Another term for "noise floor" is Minimum Discernible Signal or MDS. It is the minimum signal that can be heard above the noise generated in the receiver. Any signal smaller than this would be lost in the noise.

Noise in a receiver can be generated in the RF amplifier tubes or transistors, as well as tubes in the mixer, IF or audio amplifier stages. It can also be generated in wire conductors, resistors and other components.

4BG-4A.3 What is the theoretical minimum <u>noise floor</u> of a receiver with a 400-Hertz bandwidth?
A. -141 dBm   B. -148 dBm   C. -174 dBm   D. -180 dBm

The answer is B. The formula for Noise Power which would be equal to the theoretical minimum noise floor is:

$$Pn = KToB$$

where: Pn is the noise power
        K is Bolzmann's Constant which is equal to 1.38×10^{-23} volts/degree
        To is 290 kelvins (absolute temperature scale)
        B is the bandwidth in Hz
We then substitute the known values into the equation:

$Pn = 1.38 \times 10^{-23} \times 290 \times 400 = 160080 \times 10^{-23} = 16 \times 10^{-19}$ watts.

To express this noise floor (noise power) in dB, we compare it to a basic standard value of 1 milliwatt (.001 watts). We use the dbm power formula to do this. The "m" in dbm indicates the milliwatt standard.

$$dbm = 10 \log \frac{P2}{P1} = 10 \log \frac{16 \times 10^{-19} \text{ watts}}{.001 \text{ watts}} = 10 \log \frac{16 \times 10^{-19}}{10^{-3}} =$$

$$10 \log 16 \times 10^{-16} = 10 (\log 16 + \log 10^{-16}) = 10 (1.2 + -16) = 10 (-14.8) = -148 \text{ dbm}$$

4BG-4B.1 How can <u>selectivity</u> be achieved in the front-end circuitry of a communications receiver?
A. By using an audio filter
B. By using a preselector
C. By using an additional RF amplifier stage
D. By using an additional IF amplifier stage
    The answer is B. A preselector is a circuit that is used to improve the selectivity of a receiver. It reduces image response, intermodulation distortion, overload and cross-modulation. In many cases, it is part of the receiver's internal circuitry (and is often so identified in the receiver through a "Preselector Tuning" or similar control). The frequency range over which the preselector will tune is determined by the L-C circuit ahead of the signal amplifying circuit.

4BG-4B.2 A receiver selectivity of 2.4 kHz in the IF circuitry is optimum for what type of amateur signals?
A. CW   B. SSB voice   C. Double-sideband AM voice
D. FSK RTTY
    The answer is B. In communications work, it is only necessary to reproduce audio frequencies up to approximately 2500 Hz. A "normal" AM RF signal contains sidebands up to 2500 Hz on either side of the carrier. An SSB signal contains only one set of sidebands. Therefore, the total bandwidth of an SSB signal is approximately 2.5 kHz. The bandwidth of the IF circuitry must accommodate this 2.5 kHz bandwidth. The IF response curve should be steep enough to eliminate signals outside of the 2.5 kHz passband.

4BG-4B.3 What occurs during A1A reception if too narrow a filter bandwidth is used in the IF stage of a receiver?
A. Undesired signals will reach the audio stage
B. Output-offset overshoot
C. Cross-modulation distortion
D. Filter ringing
    The answer is D.

4BG-4B.4 What degree of selectivity is desirable in the IF circuitry of an amateur emission F1B receiver?
A. 100 Hz   B. 300 Hz   C. 6000 Hz   D. 2400 Hz

The answer is B. F1B emission is frequency-shift radioteletype. This generally uses a 170 Hz shift and a 300 Hz IF response is adequate.

4BG-4B.5 A receiver selectivity of 10 kHz in the IF circuitry is optimum for what type of amateur signals?
A. SSB voice   B. Double-sideband AM
C. CW   D. FSK RTTY

The answer is B. Double sideband AM contains sideband frequencies on both sides of the carrier. In order to accommodate this wide frequency signal, we need a wide bandpass IF circuit.

4BG-4B.6 What degree of selectivity is desirable in the IF circuitry of an emission J3E receiver?
A. 1 kHz   B. 2.4 kHz   C. 4.2 kHz   D. 4.8 kHz

The answer is B. See discussion of question 4BG-4B.2.

4BG-4B.7 What is an undesirable effect of using too wide a filter bandwidth in the IF section of a receiver?
A. Output-offset overshoot
B. Undesired signals will reach the audio stage
C. Thermal-noise distortion
D. Filter ringing

The answer is B. The IF response curve should only be wide enough to accommodate the desired signal. If it is too wide, signals on nearby frequencies, which get through the earlier stages, will get through the IF stages and reach the audio stages.

4BG-4B.8 How should the filter bandwidth of a receiver IF section compare with the bandwidth of a received signal?
A. Filter bandwidth should be slightly greater than the received-signal bandwidth
B. Filter bandwidth should be approximately half the received- signal bandwidth
C. Filter bandwidth should be approximately two times the received-signal bandwidth
D. Filter bandwidth should be approximately four times the received-signal bandwidth

The answer is A. The IF bandwidth should be slightly greater than the signal to make certain that it accommodates the entire signal and to make certain that no part of the signal is lost in the event that there is some slight detuning of the local oscillator.

4BG-4B.9 What degree of selectivity is desirable in the IF circuitry of an emission F3E receiver?
A. 1 kHz   B. 2.4 kHz   C. 4.2 kHz   D. 15 kHz

The answer is D. F3E stands for frequency modulated telephony. Narrow band FM calls for plus or minus 5 kHz deviation or a total bandwidth of 10 kHz. In actual practice, the bandwidth should extend to at least 12 kHz.

4BG-4B.10 How can selectivity be achieved in the IF circuitry of a

communications receiver?

A. Incorporate a means of varying the supply voltage to the local oscillator circuitry
B. Replace the standard JFET mixer with a bipolar transistor followed by a capacitor of the proper value
C. Remove AGC action from the IF stage and confine it to the audio stage only
D. Incorporate a high-Q filter

The answer is D. A high-Q filter improves selectivity by picking a narrow range of frequencies and sharply rejecting nearby frequency signals.

4BG-4C.1 What is meant by the <u>dynamic range</u> of a communications receiver?

A. The number of kHz between the lowest and the highest frequency to which the receiver can be tuned
B. The maximum possible undistorted audio output of the receiver, referenced to one milliwatt
C. The ratio between the minimum discernible signal and the largest tolerable signal without causing audible distortion products
D. The difference between the lowest-frequency signal and the highest-frequency signal detectable without moving the tuning knob

The answer is C. The dynamic range of a receiver is the difference between the noise floor at the low sensitivity end (where the signal is barely discernible) and the point at the high sensitivity end where the intermodulation distortion (IMD) products make reception difficult. In essence, this is the range over which the output of the receiver is a linear reproduction of the input signal. Upper and lower limits are specified and the range is measured in dB. For instance, if a certain receiver has a noise floor of -150dBm and a -80 dBm upper level, the dynamic range is 70 dB.

4BG-4C.2 What is the term for the ratio between the largest tolerable receiver input signal and the minimum discernible signal?

A. Intermodulation distortion   B. Noise floor
C. Noise figure   D. Dynamic range

The answer is D. See answer 4BG-4C.1.

4BG-4C.3 What type of problems are caused by poor <u>dynamic range</u> in a communications receiver?

A. Cross-modulation of the desired signal and desensitization from strong adjacent signals
B. Oscillator instability requiring frequent retuning, and loss of ability to recover the opposite sideband, should it be transmitted
C. Cross-modulation of the desired signal and insufficient audio power to operate the speaker
D. Oscillator instability and severe audio distortion of all but the strongest received signals

The answer is A. Cross-modulation is the superimposing or modulating of an unwanted signal onto the desired signal. Desensitization is the reduction in sensitivity due to overload from strong unwanted signals. Intermodulation distortion, and other spurious responses, are other problems caused by poor dynamic range.

4BG-4C.4 The ability of a communications receiver to perform well in the presence of strong signals outside the amateur band of interest is indicated by what parameter?
A. Noise figure
B. Blocking dynamic range
C. Signal-to-noise ratio
D. Audio output
    The answer is B. See answer 4BG-4C.3.

4BG-4D.1 What is meant by the term noise figure of a communications receiver?
A. The level of noise entering the receiver from the antenna
B. The relative strength of a received signal 3 kHz removed from the carrier frequency
C. The level of noise generated in the front end and succeeding stages of a receiver
D. The ability of a receiver to reject unwanted signals at frequencies close to the desired one
    The answer is C. The noise figure is one way of expressing the sensitivity of a receiver. Noise figure may be defined as the signal-to-noise ratio of an ideal receiver to the signal-to-noise ratio of the actual receiver. Noise figures are expressed in decibels and from the definition, we can see that the smaller the ratio, the better is the receiver sensitivity. A noise generator may be used to measure the noise figure of a receiver.

4BG-4D.2 Which stage of a receiver primarily establishes its noise figure?
A. The audio stage
B. The IF strip
C. The RF stage
D. The local oscillator
    The answer is C. The gain of the RF amplifier stage is of primary importance in determining the noise figure of the receiver. A certain amount of noise is generated in the RF amplifier stage; however, the mixer stage generates most of the receiver noise. The RF amplifier stage must, therefore, provide a strong enough signal, with low noise, to the mixer stage to overcome the mixer noise, thereby creating a high signal-to-noise ratio. As a result, the signal level at the mixer input is the limiting factor in the ability of the receiver to reproduce an acceptable output with a weak signal input from the antenna.

4BG-5A.1 What is an inverting op-amp circuit?
A. An operational amplifier circuit connected such that the input and output signals are 180 degrees out of phase
B. An operational amplifier circuit connected such that the input and output signal are in phase
C. An operational amplifier circuit connected such that the input and output are 90 degrees out of phase
D. An operational amplifier circuit connected such that the input impedance is held at zero, while the output impedance is high

The answer is A. Figure 4BG-5A.1 illustrates a basic Op-Amp inverting circuit. Note that the signal at the output is 180 degrees out of phase with the signal at the input.

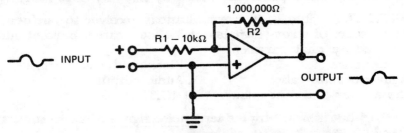

Fig. 4BG-5A.1. An inverting op-amp amplifier.

4BG-5B.1 What is a <u>noninverting op-amp circuit</u>?

A. An operational amplifier circuit connected such that the input and output signals are 180 degrees out of phase
B. An operational amplifier circuit connected such that the input and output signal are in phase
C. An operational amplifier circuit connected such that the input and output are 90 degrees out of phase
D. An operational amplifier circuit connected such that the input impedance is held at zero while the output impedance is high

The answer is B. Figure 4BG-5B.1 shows the input and output of a non-inverting Op-Amp.

Fig. 4BG-5B.1. Input and output signals of an op-amp non-inverting amplifier.

4BG-5C.1 What voltage gain can be expected from the circuit in Figure 4BG-5 when R1 is 1000 ohms and Rf is 100 kilohms?

A. 0.01
B. 1
C. 10
D. 100

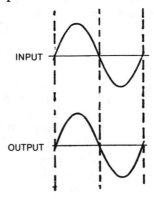

Fig 4BG-5

The answer is D. Figure 4BG-5 illustrates an inverting OP AMP.

The voltage gain is given by the following simple formula:

$$\text{GAIN} = \frac{Rf}{R_1}$$

We can then substitute the given resistance values in the equation and solve for gain:

$$\text{Gain} = \frac{100,000}{1000} = 100$$

4BG-5C.2 What voltage gain can be expected from the circuit in Figure 4BG-5 when Rl is 1800 ohms and Rf is 68 kilohms?
A. 1   B. 0.03   C. 38   D. 76
The answer is C. This problem is similar to 4BG-5C.1 and is solved in the same manner:

$$\text{Gain} = \frac{68,000}{1800} = 37.78$$

4BG-5C.3 What voltage gain can be expected from the circuit in Figure 4BG-5 when Rl is 3300 ohms and Rf is 47 kilohms?
A. 28   B. 14   C. 7   D. 0.07
The answer is B. This problem is similar to 4BG-5C.1 and is solved in the same manner:

$$\text{Gain} = \frac{47,000}{3300} = 14.24$$

4BG-5C.4 What voltage gain can be expected from the circuit in Figure 4BG-5 when Rl is 10 ohms and Rf is 47 kilohms?
A. 0.00021   B. 9400   C. 4700   D. 2350
The answer is C. This problem is similar to 4BG-5C.1 and is solved in the same manner:

$$\text{Gain} = \frac{47,000}{10} = 4700$$

4BG-5D.1 How does the gain of a theoretically ideal operational amplifier vary with frequency?
A. The gain increases linearly with increasing frequency
B. The gain decreases linearly with increasing frequency
C. The gain decreases logarithmically with increasing frequency
D. The gain does not vary with frequency
The answer is D. See questions 4BF-2.1 and 4BF-2.3.

4BG-6.1 What determines the input impedance in a FET common-source amplifier?
A. The input impedance is essentially determined by the resistance between the drain and substrate
B. The input impedance is essentially determined by the resistance

between the source and drain
C. The input impedance is essentially determined by the gate biasing network
D. The input impedance is essentially determined by the resistance between the source and substrate

The answer is C. In the case of a JFET, the gate is always reverse biased. Thus, the input resistance is very high. In the case of a MOSFET, the input impedance is very high because the gate is insulated from the rest of the device. Thus, the input impedance is determined by the resistance which is in parallel or across the input. This is shown in figure 4BG-6.1.

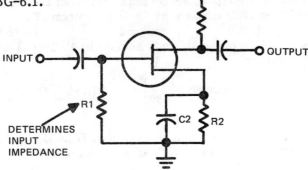

Fig. 4BG-6.1. An FET common source amplifier.

4BG-6.2 What determines the output impedance in a FET common-source amplifier?
A. The output impedance is essentially determined by the drain resistor
B. The output impedance is essentially determined by the input impedance of the FET
C. The output impedance is essentially determined by the drain-supply voltage
D. The output impedance is essentially determined by the gate supply voltage

The answer is A. The drain of an FET is similar to the plate of a tube. The drain resistor is across the output and therefore determines, to a large degree, the output impedance of the amplifier.

4BG-7.1 What frequency range will be tuned by the circuit in Figure 4BG-7 when L is 10 microhenrys, Cf is 156 picofarads, and Cv is 50 picofarads maximum and 2 picofarads minimum?
A. 3508 through 4004 kHz
B. 6998 through 7360 kHz
C. 13.396 through 14.402 MHz
D. 49.998 through 54.101 MHz

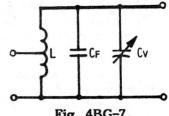

Fig. 4BG-7.

The answer is A. Cf and Cv are in parallel. Therefore, the total capacity is the sum of the two capacitors. The total capacity at the low frequency end of the range is 156 pf + 50 pf or 206 pf. The total capacity at the high frequency end of the range is 156 pf + 2 pf or 158 pf. We substitute these two values in the resonance formula to arrive at the tuning range.

$$\text{Resonant frequency} = \frac{10^6}{2\pi\sqrt{LC}} \quad \text{where L is in microhenries and C is in picofarads}$$

$$\text{f res (low)} = \frac{10^6}{6.28\sqrt{10 \times 206}} = \frac{10^6}{285.03} = 3508.4 \text{ kHz}$$

$$\text{f res (high)} = \frac{10^6}{6.28\sqrt{10 \times 158}} = \frac{10^6}{249.625} = 4006 \text{ kHz}$$

4BG-7.2 What frequency range will be tuned by the circuit in Figure 4BG-7 when L is 30 microhenrys, Cf is 200 picofarads, and Cv is 80 picofarads maximum and 10 picofarads minimum?
A. 1737 through 2005 kHz   B. 3507 through 4004 kHz
C. 7002 through 7354 kHz   D. 14.990 through 15.020 MHz
The answer is A. We use the same reasoning and method to solve this problem as in 4BG-7.1:

$$\text{f res (low)} = \frac{10^6}{6.28\sqrt{30 \times 280}} = \frac{10^6}{575.57} = 1737.4 \text{ kHz}$$

$$\text{f res (high)} = \frac{10^6}{6.28\sqrt{30 \times 210}} = \frac{10^6}{498.46} = 2006 \text{ kHz}$$

4BG-8.1 What is the purpose of a bypass capacitor?
A. It increases the resonant frequency of the circuit
B. It removes direct current from the circuit by shunting dc to ground
C. It removes alternating current by providing a low impedance path to ground
D. It acts as a voltage divider
The answer is C. The purpose of a by-pass capacitor is to get rid of the AC component of a signal at a certain point. It will not affect the DC component because a capacitor acts as an open circuit for DC. In figure 4BG-8.1, C1 is a by-pass capacitor. Its purpose is to keep the voltage across R1 constant. C1 by-passes the AC component "around" R1 to ground.

4BG-8.2 What is the purpose of a coupling capacitor?
A. It blocks direct current and passes alternating current
B. It blocks alternating current and passes direct current
C. It increases the resonant frequency of the circuit
D. It decreases the resonant frequency of the circuit
The answer is A. A coupling capacitor is used where we want to block the DC component of a signal and pass the AC component. C2, in Figure 4BG-8.1, blocks the DC at the collector and prevents

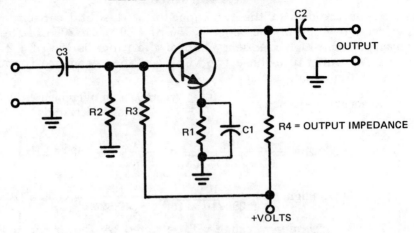

Fig. 4BG–8.1. A Common emitter amplifier.

it from getting on to the next stage, while at the same time, passing the AC component of the signal.

SUBELEMENT 4BH
SIGNALS AND EMISSIONS
(4 questions)

4BH-1A.1 In a pulse-width modulation system, what parameter does the modulating signal vary?
A. Pulse duration
B. Pulse frequency
C. Pulse amplitude
D. Pulse intensity

The answer is A. Pulse modulation is a general term used for transmitting intelligence by discrete carrier pulses instead of a continuous waveform. The position, width or amplitude of the pulses can be altered to convey intelligence.

In the pulse-width modulation system (also known as pulse duration modulation), all pulses are of equal amplitude and their relative positions to each other are the same. However, the pulses are of different durations (or widths). The exact amount of the duration depends upon the waveform of the modulating signal at that instant. The pulse duration is usually on the order of milliseconds or less. See Figure 4BH-1A.1.

Fig. 4BH-1A.1. Pulse-width modulation.

4BH-1A.2 What is the type of modulation in which the modulating signal varies the duration of the transmitted pulse?
A. Amplitude modulation
B. Frequency modulation
C. Pulse-width modulation
D. Pulse-height modulation

The answer is C. See answer 4BH-1A.1.

4BH-1B.1 In a pulse-position modulation system, what parameter does the modulating signal vary?
A. The number of pulses per second
B. Both the frequency and amplitude of the pulses
C. The duration of the pulses
D. The time at which each pulse occurs

The answer is D. Pulse-Position modulation uses pulses of the same amplitude and duration, but with different positions with respect to each other. The spacing between the pulses varies. Information is conveyed by the varying spacing. See Figure 4BH-1B.1. Current FCC rules prohibit both pulse-width and pulse-position modulation on frequencies below 2300 MHz.

Fig. 4BH-1B.1. Pulse-position modulation.

4BH-1B.2 Why is the transmitter peak power in a pulse modulation

system much greater than its average power?

A. The signal duty cycle is less than 100%
B. The signal reaches peak amplitude only when voice-modulated
C. The signal reaches peak amplitude only when voltage spikes are generated within the modulator
D. The signal reaches peak amplitude only when the pulses are also amplitude-modulated

The answer is A. From Figures 4BH-1A.1 and 4BH-1B.1, we can see that the signal reaches its peak for only a short time, compared to the time where the signal is zero. Therefore, if we consider the "peaks" and the "zeros", the average power is much less than the peak power.

4BH-1B.3 What is one way that voice is transmitted in a pulse- width modulation system?

A. A standard pulse is varied in amplitude by an amount depending on the voice waveform at that instant
B. The position of a standard pulse is varied by an amount depending on the voice waveform at that instant
C. A standard pulse is varied in duration by an amount depending on the voice waveform at that instant
D. The number of standard pulses per second varies depending on the voice waveform at that instant

The answer is C. See question 4BH-1A.1.

4BH-2A.1 What digital code consists of elements having unequal length?

A. ASCII   B. AX.25   C. Baudot   D. Morse code

The answer is D. The Morse Code is a digital code. When the telegraph key is pressed down we get our "high" state of full power out. When the key is opened, we get our "low" state of zero power out. In addition, the unequal lengths of the "high" state gives us a full code of alphabet and numbers.

4BH-2B.1 What digital communications system is well suited for meteor-scatter communications?

A. ACSSB   B. AMTOR   C. Packet radio D. Spread spectrum

The answer is C. Meteor-scatter communications require short transmissions. It is important to get as much information as possible into a short period of time. Since Packet Radio sends burst of data at high speeds, it is ideally suited for meteor-scatter communications.

4BH-2B.2 The International Organization for Standardization has developed a seven-level reference model for a packet-radio communications structure. What level is responsible for the actual transmission of data and handshaking signals?

A. The physical layer   B. The transport layer
C. The communications layer   D. The synchronization layer

The answer is A. The International Organization for Standardization has developed a reference model consisting of seven levels or layers. They are; 1. Physical layer, 2. Link layer, 3. Network layer, 4.

Transport layer, 5. Session layer, 6. Presentation layer and 7. Application layer.

The physical layer deals with the binary state of ones and zeros, modems and similar equipment.

4BH-2B.3 The International Organization for Standardization has developed a seven-level reference model for a packet-radio communications structure. What level arranges the bits into frames and controls data flow?

A. The transport layer   B. The link layer
C. The communications layer   D. The synchronization layer

The answer is B. See question 4BH-2B.2. The Link layer does error detection.

4BH-2C.1 What is one advantage of using the ASCII code, with its larger character set, instead of the Baudot code?

A. ASCII includes built-in error-correction features
B. ASCII characters contain fewer information bits than Baudot characters
C. It is possible to transmit upper and lower case text
D. The larger character set allows store-and-forward control characters to be added to a message

The answer is C. The ASCII code uses seven information pulses or bits to represent letters, numbers, etc. The Baudot code uses only five. Thus, the ASCII code has more possible combinations than the Baudot code and this allows the ASCII code to have upper and lower case letters of the alphabet.

4BH-2D.1 What type of error control system does <u>Mode A AMTOR</u> use?

A. Each character is sent twice
B. The receiving station checks the calculated frame check sequence (FCS) against the transmitted FCS
C. Mode A AMTOR does not include an error control system
D. The receiving station automatically requests repeats when needed

The answer is D. AMTOR is a system used in RTTY to improve reception. It does so by sending the same signal at two different times. Thus, if anything is missed the first time, it can be heard the second time. This is known as TIME DIVERSITY.

There are two different methods that AMTOR uses to accomplish the principle of TIME DIVERSITY. In one system, MODE A, the information is repeated only when the receiving station requests the information. In the second method, Mode B, the signal is automatically sent twice.

4BH-2D.2 What type of error control system does <u>Mode B AMTOR</u> use?

A. Each character is sent twice
B. The receiving station checks the calculated frame check sequence (FCS) against the transmitted FCS
C. Mode B AMTOR does not include an error control system

D. The receiving station automatically requests repeats when needed
The answer is A. See question 4BH-2D.1.

4BH-2E.1 What is the duration of a 45-baud Baudot RTTY data pulse?

A. 11 milliseconds   B. 40 milliseconds
C. 31 milliseconds   D. 22 milliseconds

The answer is D. Baudot is a code used in radioteletype. In the Baudot code, each character is made up of five data elements plus a start element and a stop element. An element is either a space (0) or a mark (1). The different characters (letters of the alphabet, numbers, etc.) have different combinations of spaces and marks. Figure 4BH-2E.1A shows the letter "J". The five data pulses that characterize the letter J are: mark, mark, space, mark, space. The five data pulses for Y are: mark, space, mark, space, mark. In front of the five data pulses, there is a start pulse which is always a space pulse. After the five data pulses, there is a stop pulse which is always a mark pulse.

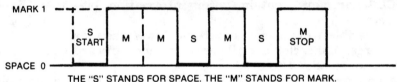

THE "S" STANDS FOR SPACE. THE "M" STANDS FOR MARK.

Fig. 4BH-2E.1A. The letter J in Baudot code.

The baud is a unit of signaling speed. It is equal to the number of signal events in one second We can state this mathematically as follows:

$$T = \frac{1}{B} \quad \text{or} \quad B = \frac{1}{T}$$

where: T is the time duration of the bit in seconds
           B is the Baud rate

We solve the question by substituting the given Baud rate in the first formula:

$$T = \frac{1}{B} = \frac{1}{45} = 0.022 \text{ seconds or 22 microseconds}$$

In common practice, 45-baud Baudot RTTY has a data pulse time duration of 22 ms. (milliseconds) and a start pulse time duration of 22 ms. However, the stop pulse duration is 31 ms. Higher RTTY pulse speeds have shorter time duration pulses.

4BH-2E.2 What is the duration of a 45-baud Baudot RTTY start pulse?

A. 11 milliseconds   B. 22 milliseconds
C. 31 milliseconds   D. 40 milliseconds

The answer is B See question 4BH-2E.1.

4BH-2E.3 What is the duration of a 45-baud Baudot stop pulse?

A. 11 milliseconds   B. 18 milliseconds
C. 31 milliseconds   D. 40 milliseconds
   The answer is C. See question 4BH-2E.1.

4BH-2E.4 What is the primary advantage of AMTOR over Baudot RTTY?
A. AMTOR characters contain fewer information bits than Baudot characters
B. AMTOR includes an error detection system
C. Surplus radioteletype machines that use the AMTOR code are readily available
D. Photographs can be transmitted using AMTOR
   The answer is B. See question 4BH-2D.1.

4BH-2F.1 What is the necessary bandwidth of a 170-Hertz shift, 45-baud Baudot emission F1B transmission?
A. 45 Hz   B. 250 Hz   C. 442 Hz   D. 600 Hz
   The answer is B. We use the following formula to find the bandwidth:

$$\text{Bandwidth} = \text{Baud rate} + (1.2 \times \text{Frequency shift})$$

$$\text{Bandwidth} = 45 + (1.2 \times 170) = 45 + 204 = 249 \text{ Hz}$$

4BH-2F.2 What is the necessary bandwidth of a 170-Hertz shift, 45-baud Baudot emission J2B transmission?
A. 45 Hz   B. 249 Hz   C. 442 Hz   D. 600 Hz
   The answer is B. See answer 4BH-2F.1. Since the baud rate and frequency shift are the same as in question 4BH-2F.1, we arrive at the same bandwidth when substituting the same values in the formula.

4BH-2F.3 What is the necessary bandwidth of a 170-Hertz shift, 74-baud Baudot emission F1B transmission?
A. 250 Hz   B. 278 Hz   C. 442 Hz   D. 600 Hz
   The answer is B. We use the same method to solve this problem as we did in answer 4BH-2F.1.

$$\text{Bandwidth} = 74 + (1.2 \times 170) = 74 + 204 = 278 \text{ Hz}$$

4BH-2F.4 What is the necessary bandwidth of a 170-Hertz shift, 74-baud Baudot emission J2B transmission?
A. 250 Hz   B. 278 Hz   C. 442 Hz   D. 600 Hz
   The answer is B. See answer 4BH-2F.3. Since the baud rate and frequency shift are the same as in question 4BH-2F.3, we arrive at the same bandwidth when substituting the same values in the formula.

4BH-2F.5 What is the necessary bandwidth of a 13-WPM international Morse code emission A1A transmission?
A. Approximately 13 Hz   B. Approximately 26 Hz
C. Approximately 52 Hz   D. Approximately 104 Hz
   The answer is C. In order to find the bandwidth of a CW signal, we use the following formula:

BW = BK   where: BW is the bandwidth in Hz
                  B is the baud rate which is .8 X wpm
                  K is a fading factor that is equal to 5

We substitute the values given in the problem into the formula:

Bandwidth = .8 X 13 X 5 = 10.4 X 5 = 52 Hz

4BH-2F.6 What is the necessary bandwidth of a 13-WPM international Morse code emission J2A transmission?
A. Approximately 13 Hz   B. Approximately 26 Hz
C. Approximately 52 H   D. Approximately 104 Hz
    The answer is C. This problem is solved in the same manner as answer 4BH-2F.5.

Bandwidth = .8 X 13 X 5 = 52 Hz

4BH-2F.7 What is the necessary bandwidth of a 1000-Hertz shift, 1200-baud ASCII emission F1D transmission?
A. 1000 Hz   B. 1200 Hz   C. 440 Hz   D. 2400 Hz
    The answer is D. We use the same formula to solve this problem we did in answer 4BH-2F.1.

Bandwidth = Baud rate + (1.2 X frequency shift)

Bandwidth = 1200 + (1.2 X 1000) = 1200 + 1200 = 2400 Hz

4BH-2F.8 What is the necessary bandwidth of a 4800-hertz frequency shift, 9600-baud ASCII emission F1D transmission?
A. 15.36 kHz   B. 9.6 kHz   C. 4.8 kHz   D. 5.76 kHz
    The answer is A. This problem is similar to question 4BH-2F.7 and is solved in the same manner.

Bandwidth = 9600 + (1.2 X 4800) = 9600 + 5760 = 15360 Hz

15360 Hz = 15.36 kHz

4BH-2F.9 What is the necessary bandwidth of a 4800-hertz frequency shift, 9600-baud ASCII emission J2D transmission?
A. 15.36 kHz   B. 9.6 kHz   C. 4.8 kHz   D. 5.76 kHz
    The answer is A. This problem is solved in the same manner as answer 4BH-2F.8. The answer is the same because the given values are the same.

4BH-2F.10 What is the necessary bandwidth of a 5-WPM international Morse code emission A1A transmission?
A. Approximately 5 Hz   B. Approximately 10 Hz
C. Approximately 20 Hz   D. Approximately 40 Hz
    The answer is C. This problem is solved in exactly the same manner as answer 4BH-2F.5.

Bandwidth = .8 X 5 X 5 = 4 X 5 = 20 Hz

4BH-2F.11 What is the necessary bandwidth of a 5-WPM international Morse code emission J2A transmission?

A. Approximately 5 Hz   B. Approximately 10 Hz
C. Approximately 20 Hz   D. Approximately 40 Hz

The answer is C. This problem is solved in the same manner as answer 4BH-2F.10. The answer is the same because the given values are the same.

4BH-2F.12 What is the necessary bandwidth of a 170-Hertz shift, 110-baud ASCII emission F1B transmission?
A. 304 Hz   B. 314 Hz   C. 608 Hz   D. 628 Hz

The answer is B. We use the same formula to solve this problem as we did in answer 4BH-2F.1.

$$\text{Bandwidth} = 110 + (1.2 \times 170) = 110 + 204 = 314 \text{ Hz}$$

4BH-2F.13 What is the necessary bandwidth of a 170-Hertz shift, 110-baud ASCII emission J2B transmission?
A. 304 Hz   B. 314 Hz   C. 608 Hz   D. 628 Hz

The answer is B. This problem is solved in the same manner as answer 4BH-2F.12. The answer is the same because the given values are the same.

4BH-2F.14 What is the necessary bandwidth of a 170-Hertz shift, 300-baud ASCII emission F1D transmission?
A. 0 Hz   B. 0.3 kHz   C. 0.5 kHz   D. 1.0 kHz

The answer is C. We use the same method to solve this problem as we did in answer 4BH-2F.12.

$$\text{Bandwidth} = 300 + (1.2 \times 170) = 300 + 204 = 504 \text{ Hz}$$
$$504 \text{ Hz} = 0.5 \text{ kHz}$$

4BH-2F.15 What is the necessary bandwidth for a 170-Hertz shift, 300-baud ASCII emission J2D transmission?
A. 0 Hz   B. 0.3 kHz   C. 0.5 kHz   D. 1.0 kHz

The answer is C. This problem is solved in exactly the same manner as answer 4BH-2F.14. The answer is the same because the given values are the same.

4BH-3.1 What is <u>amplitude compandored single sideband?</u>
A. Reception of single sideband with a conventional CW receiver
B. Reception of single sideband with a conventional FM receiver
C. Single sideband incorporating speech compression at the transmitter and speech expansion at the receiver
D. Single sideband incorporating speech expansion at the transmitter and speech compression at the receiver

The answer is C. There are two types of compandoring. One is amplitude compandoring; the other is frequency compandoring. In amplitude compandoring, a circuit in the transmitter increases the amplitude of the weaker portions of speech and decreases the amplitude of the stronger components, thereby compressing the signal into a narrow amplitude range. The result is a voice signal with an almost constant average peak signal. This, in turn, results in a narrower bandwidth signal which causes less interference to other signals and permits a sharper bandpass at the receiving end. A narrower receiver bandpass means less interference from adjacent signals. At the receiving end, the receiver has a circuit that expands and thereby restores the signal's original amplitude variations. The term "compandor" comes about by combining part of the word COMpress and part of the word exPANDOR.

In frequency compandoring, a circuit in the transmitter changes the higher frequency components into lower frequencies, resulting in a signal with a narrower bandwidth. At the receiver, the higher frequencies are "expandored" to their original values, thereby restoring the fidelity of the original signal.

4BH-3.2 What is meant by compandoring?
A. Compressing speech at the transmitter and expanding it at the receiver
B. Using an audio-frequency signal to produce pulse-length modulation
C. Combining amplitude and frequency modulation to produce a single-sideband signal
D. Detecting and demodulating a single-sideband signal by converting it to a pulse-modulated signal
 The answer is A. See answer 4BH-3.1.

4BH-3.3 What is the purpose of a pilot tone in an amplitude compandored single sideband system?
A. It permits rapid tuning of a mobile receiver
B. It replaces the suppressed carrier at the receiver
C. It permits rapid change of frequency to escape high-powered interference
D. It acts as a beacon to indicate the present propagation characteristic of the band
 The answer is A. In "normal" single sideband, the carrier is suppressed at the transmitter and the sidebands are transmitted. In order to detect the signal at the receiver, a carrier is generated and reinserted into the signal. However, there is no guarantee that this new carrier will have the exact correct frequency, and tuning becomes somewhat difficult and slow.

In order to overcome this problem, a low volume audio tone of approximately 3 kHz, called a "pilot tone", is added to the signal at the transmitter. At the receiver, the pilot tone is used to lock-in the receiver's local oscillator to the correct frequency though a phase-locked loop circuit. The pilot tone is filtered out so that it does not interfere with the audio.

4BH-3.4 What is the approximate frequency of the pilot tone in an amplitude compandored single sideband system?
A. 1 kHz   B. 5 MHz   C. 455 kHz   D. 3 kHz
 The answer is D. See answer 4BH-3.3.

4BH-3.5 How many more voice transmissions can be packed into a given frequency band for amplitude compandored single sideband systems over conventional emission F3E systems?

A. 2   B. 4   C. 8   D. 16

The answer is B. The narrow band FM used by amateurs has a frequency deviation of + or - 5 kHz or a total frequency swing of 10 kHz. This translates to a bandwidth of approximately 12 to 15 kHz. Amplitude compandored single sideband uses only one set of sidebands, giving us a maximum bandwidth of approximately 3 kHz. Allowing for space between signals, we can therefore have about four times as many SSB signals as FM signals.

4BH-4.1 What term describes a wide-bandwidth communications system in which the RF carrier varies according to some predetermined sequence?
A. Amplitude compandored single sideband
B. AMTOR
C. Time-domain frequency modulation
D. Spread spectrum communication

The answer is D. In spread spectrum (SS) communication, the signal is spread out over a wide frequency band by special coded modulation techniques. At the receiver, the same code is used to narrow the signal down to a "normal" width. SS communications has the advantages of improved signal-to-noise ratio and less interference.

4BH-4.2 What is the term used to describe a spread spectrum communications system where the center frequency of a conventional carrier is altered many times per second in accordance with a pseudo-random list of channels?
A. Frequency hopping
B. Direct sequence
C. Time-domain frequency modulation
D. Frequency compandored spread spectrum

The answer is A. In the "Frequency hopping" system, the RF output is emitted on different frequencies in accordance with a prearranged sequence. It stays at each frequency for a very short time.

4BH-4.3 What term is used to describe a spread spectrum communications system in which a very fast binary bit stream is used to shift the phase of an RF carrier?
A. Frequency hopping
B. Direct sequence
C. Binary phase-shift keying
D. Phase compandored spread spectrum

The answer is B. In "Direct sequence", a high speed coded signal modulates a carrier. A few of the other terms used for "Direct sequence" are phase hopping, psendo noise and direct spread.

4BH-5.1 What is the term for the amplitude of the maximum positive excursion of a signal as viewed on an oscilloscope?
A. Peak-to-peak voltage   B. Inverse peak negative voltage
C. RMS voltage   D. Peak positive voltage

The answer is D. Figure 4BH-5.1 illustrates the graph of a cycle of an AC sine wave signal. Note that everything above the 0 horizontal line is considered positive (+), and everything below the 0 horizontal line is considered negative (-). The peak positive voltage is from the zero (0) line to the highest point of the wave in the positive direction.

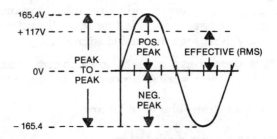

Fig. 4BH–5.1. An AC sine wave.

4BH–5.2 What is the term for the amplitude of the maximum negative excursion of a signal as viewed on an oscilloscope?
A. Peak-to-peak voltage   B. Inverse peak positive voltage
C. RMS voltage   D. Peak negative voltage
     The answer is D. See answer and Figure 4BH–5.1. The peak negative voltage is from the zero (0) line to the highest point of the wave in the negative direction.

4BH–6A.1 What is the easiest voltage amplitude dimension to measure by viewing a pure sine wave signal on an oscilloscope?
A. Peak-to-peak voltage   B. RMS voltage
C. Average voltage   D. DC voltage
     The answer is A. This can be seen from Figure 4BH–5.1. A transparent grid can be placed over the face of the cathode ray tube. It is then a simple matter to measure the peak-to-peak voltage.

4BH–6A.2 What is the relationship between the peak-to-peak voltage and the peak voltage amplitude in a symmetrical wave form?
A. 1:1   B. 2:1   C. 3:1   D. 4:1
     The answer is B. See Figure 4BH–5.1. When we say symmetrical, we mean that the amplitude of the signal above the horizontal line is equal to the amplitude of the signal below the line. By examining Figure 4BH–5.1, which is a symmetrical wave, it is easy to see that the peak-to-peak voltage is twice the peak (negative or positive) voltage.

4BH–6A.3 What input–amplitude parameter is valuable in evaluating the signal-handling capability of a Class A amplifier?
A. Peak voltage   B. Average voltage
C. RMS voltage   D. Resting voltage
     The answer is A. In evaluating the signal handling capability of a linear device, we need to know how it will react to maximum or "peak" conditions.

SUBELEMENT 4BI
ANTENNAS AND FEEDLINES
(4 questions)

4BI-1A.l What is an <u>isotropic radiator</u>?
A. A hypothetical, omnidirectional antenna
B. In the northern hemisphere, an antenna whose directive pattern is constant in southern directions
C. An antenna high enough in the air that its directive pattern is substantially unaffected by the ground beneath it
D. An antenna whose directive pattern is substantially unaffected by the spacing of the elements

The answer is A. An isotropic radiator is a theoretically perfect antenna that radiates equally well in all directions. It is a standard reference antenna, and the gain of other antennas are compared to it. For instance, when we say that a particular Yagi antenna has a gain of 7 dB, we mean 7 dB more than the isotropic reference antenna.

It should be noted that there are two types of reference antennas that are used for gain comparison. The isotropic antenna is one of them. The other is a simple dipole antenna. The simple dipole antenna standard has a gain of 2.1 dB, compared to the isotropic standard. Because of this difference in reference antennas, it is important that the dB gain figure for a directive antenna include the standard of reference that was used.

4BI-1B.1 When is it useful to refer to an <u>isotropic radiator</u>?
A. When comparing the gains of directional antennas
B. When testing a transmission line for standing wave ratio
C. When (in the northern hemisphere) directing the transmission in a southerly direction
D. When using a dummy load to tune a transmitter

The answer is A. See answer 4BI-1A.1.

4BI-1B.2 What theoretical reference antenna provides a comparison for antenna measurements?
A. Quarter-wave vertical   B. Yagi
C. Bobtail curtain   D. Isotropic radiator

The answer is D. See answer 4BI-1A.1.

4BI-1B.3 What purpose does an <u>isotropic radiator</u> serve?
A. It is used to compare signal strengths (at a distant point) of different transmitters
B. It is used as a reference for antenna gain measurements
C. It is used as a dummy load for tuning transmitters
D. It is used to measure the standing-wave-ratio on a transmission line

The answer is B. It is the theoretical standard or reference that the gain of other antennas are compared to. See answer 4BI-1A.1.

4BI-1B.4 How much gain does a 1/2-wavelength dipole have over an isotropic radiator?
A. About 1.5 dB B. About 2.1 dB C. About 3.0 dB D. About 6.0 dB
   The answer is B. See answer 4BI-1A.1.

4BI-1B.5 How much gain does an antenna have over a 1/2-wavelength dipole when it has 6 dB gain over an isotropic radiator?
A. About 3.9 dB B. About 6.0 dB   C. About 8.1 dB D. About 10.0 dB
   The answer is A. If a certain antenna has a 6 dB gain over an isotropic antenna, and the half wave dipole has a 2.1 dB gain over an isotropic antenna, then the difference between the certain antennas's gain of 6 dB and the dipole gain of 2.1 dB is the gain of the certain antenna over the dipole antenna.

$$6 \text{ dB} - 2.1 \text{ dB} = 3.9 \text{ dB}$$

4BI-1B.6 How much gain does an antenna have over a 1/2-wavelength dipole when it has 12 dB gain over an isotropic radiator?
A. About 6.1 dB   B. About 9.9 dB
C. About 12.0 dB   D. About 14.1 dB
   The answer is B. See answer 4BI-1B.5.

$$12 \text{ dB} - 2.1 \text{ dB} = 9.9 \text{ dB}$$

4BI-1C.1 What is the antenna pattern for an isotropic radiator?
A. A figure-8   B. A unidirectional cardioid
C. A parabola   D. A sphere
   The answer is D. Since the isotropic radiator radiates equally well in all directions, a sphere would be its pattern.

4BI-1C.2 What type of directivity pattern does an isotropic radiator have?
A. A figure-8   B. A unidirectional cardioid
C. A parabola   D. A sphere
   The answer is D. See answer 4BI-1C.1.

4BI-2A.1 What is the radiation pattern of two 1/4-wavelength vertical antennas spaced 1/2 wavelength apart and fed 180 degrees out of phase?
A. Unidirectional cardioid   B. Omnidirectional
C. Figure-8 broadside to the antennas
D. Figure-8 end-fire in line with the antennas
   The answer is D. The use of two or more vertical antennas results in better gain and directivity than one vertical antenna. By changing the spacing between the antennas and changing the phases of the currents in the antennas, various radiation patterns can be obtained.
   When two 1/4 wavelength vertical antennas are spaced 1/2 wavelength apart, and they are fed 180 degrees out of phase, the resulting radiation pattern is a figure 8 pattern. The pattern is in line with the plane of the antennas. Another way of stating this is that the maximum radiation is parallel to a line joining the two antennas. This pattern is shown in Figure 4BI-2A.1.

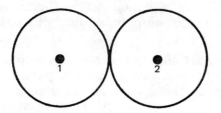

Fig. 4BI–2A.1. Figure 8 radiation pattern.

4BI–2A.2 What is the radiation pattern of two 1/4–wavelength vertical antennas spaced 1/4 wavelength apart and fed 90 degrees out of phase?
A. Unidirectional cardioid   B. Figure–8 end–fire
C. Figure–8 broadside   D. Omnidirectional
    The answer is A. Figure 4BI–2A.2 shows the resulting pattern when two 1/4 wavelength vertical antennas are spaced 1/4 wavelength apart, and fed 90 degrees out of phase. The radiation pattern is unidirectional; note the sharp null at antenna 1 and the maximum at antenna 2, in the direction indicated. The term "cardioid" is used because the radiation pattern resembles the shape of a heart.

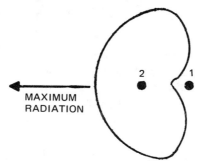

Fig. 4BI–2A.2. A cardiod radiation pattern.

4BI–2A.3 What is the radiation pattern of two 1/4–wavelength vertical antennas spaced 1/2 wavelength apart and fed in phase?
A. Omnidirectional   B. Cardioid unidirectional
C. Figure–8 broadside to the antennas
D. Figure–8 end–fire in line with the antennas
    The answer is C. Figure 4BI–2A.3 shows the directional pattern when the antennas are spaced 1/2 wavelength apart and fed in phase.

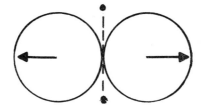

Fig. 4BI–2A.3. A Figure 8 radiation pattern.

Note that the maximum radiation of the figure 8 pattern is perpendicular (broadside) to the plane (a line joining the two antennas) of the antennas.

4BI-2A.4 How far apart should two 1/4-wavelength vertical antennas be spaced in order to produce a figure-8 pattern that is broadside to the plane of the verticals when fed in phase?
A. 1/8 wavelength   B. 1/4 wavelength
C. 1/2 wavelength   D. 1 wavelength
     The answer is C. See answer 4BI-2A.3. This question is actually a rephrasing of question 4BI-2A.3.

4BI-2A.5 How many 1/2 wavelengths apart should two 1/4-wavelength vertical antennas be spaced to produce a figure-8 pattern that is in line with the vertical antennas when they are fed 180 degrees out of phase?
A. One half wavelength apart   B. Two half wavelengths apart
C. Three half wavelengths apart   D. Four half wavelengths apart
     The answer is A. See answer 4BI-2A.1. This question is actually a rephrasing of question 4BI-2A.1.

4BI-2A.6 What is the radiation pattern of two 1/4-wavelength vertical antennas spaced 1/4 wavelength apart and fed 180 degrees out of phase?
A. Omnidirectional   B. Cardioid unidirectional
C. Figure-8 broadside to the antennas
D. Figure-8 end-fire in line with the antennas
     The answer is D. This results in a figure 8 pattern with the maximum radiation parallel to the plane of the antennas, similar to Figure 4BI-2A.1.

4BI-2A.7 What is the radiation pattern for two 1/4-wavelength vertical antennas spaced 1/8 wavelength apart and fed 180 degrees out of phase?
A. Omnidirectional
B. Cardioid unidirectional
C. Figure-8 broadside to the antennas
D. Figure-8 end-fire in line with the antennas
     The answer is D. This results in a figure 8 pattern with the maximum radiation parallel to the plane of the antennas, similar to Figure 4BI-2A.1.

4BI-2A.8 What is the radiation pattern for two 1/4-wavelength vertical antennas spaced 1/8 wavelength apart and fed in phase?
A. Omnidirectional
B. Cardioid unidirectional
C. Figure-8 broadside to the antennas
D. Figure-8 end-fire in line with the antennas
     The answer is A. Two 1/4 wavelength vertical antennas, spaced 1/8 wavelength apart and fed in phase, will produce an omnidirectional pattern. Omnidirectional means equal radiation in all directions. This pattern is represented by a circle, as shown in Figure 4BI-2A.8.

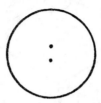

Fig. 4BI-2A.8. An omnidirectional pattern.

4BI-2A.9 What is the radiation pattern for two 1/4-wavelength vertical antennas spaced 1/4 wavelength apart and fed in phase?
A. Substantially unidirectional   B. Elliptical
C. Cardioid unidirectional
D. Figure-8 end-fire in line with the antennas

The answer is B. Two vertical antennas, spaced 1/4 wavelength apart and fed in phase, produce an elliptical pattern, as shown in Figure 4BI-2A.9. The difference between this pattern and the figure 8 pattern is that the elliptical pattern shows some radiation in directions where the figure 8 pattern shows nulls.

Fig. 4BI-2A.9. An eliptical pattern.

4BI-3A.1 What is a resonant rhombic antenna?
A. A unidirectional antenna, each of whose sides is equal to half a wavelength and which is terminated in a resistance equal to its characteristic impedance
B. A bidirectional antenna open at the end opposite that to which the transmission line is connected and with each side approximately equal to one wavelength
C. An antenna with an LC network at each vertex (other than that to which the transmission line is connected) tuned to resonate at the operating frequency
D. A high-frequency antenna, each of whose sides contains traps for changing the resonance to match the band in use

The answer is B. A rhombic antenna is basically two V-beam antennas, placed back to back in the form of a diamond. All sides have the same length. See figure 4BI-3A.1. If angle C (opposite the feedpoint) is left open, the antenna is called a resonant rhombic antenna. Its radiation pattern is bidirectional. See question 4BI-3B.1.

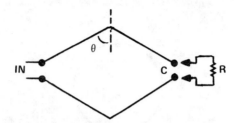

Fig. 4BI-3A.1. A rhombic antenna.

4BI-3B.1 What is a nonresonant rhombic antenna?

A. A unidirectional antenna terminated in a resistance equal to its characteristic impedance
B. An open-ended bidirectional antenna
C. An antenna resonant at approximately double the frequency of the intended band of operation
D. A horizontal triangular antenna consisting of two adjacent sides and the long diagonal of a resonant rhombic antenna

The answer is A. See Figure 4BI-3A.1. When we terminate the antenna at angle C (the end opposite the feed point) with a resistance of suitable value, the antenna is known as a non-resonant rhombic antenna. Its radiation pattern becomes unidirectional. The resistor or resistors used to terminate the antenna, should be non-inductive and have a resistance of approximately 800 ohms. The input impedance to the rhombic antenna is approximately 600 to 800 ohms.

4BI-3B.2 What are the advantages of a nonresonant rhombic antenna?

A. Wide frequency range, high gain and high front-to-back ratio
B. High front-to-back ratio, compact size and high gain
C. Unidirectional radiation pattern, high gain and compact size
D. Bidirectional radiation pattern, high gain and wide frequency range

The answer is A. Because it works well over a wide frequency range, it can be used as a multiband antenna. Another advantage of a nonresonant rhombic antenna is its low angle of radiation.

4BI-3B.3 What are the disadvantages of a nonresonant rhombic antenna?

A. It requires a large area for proper installation and has a narrow bandwidth
B. It requires a large area for proper installation and has a low front-to-back ratio
C. It requires a large amount of aluminum tubing and has a low front-to-back ratio
D. It requires a large area and four sturdy supports for proper installation

The answer is D. The gain of the rhombic antenna is greater when its wire size is greater. This calls for a large, extensive installation. Other disadvantages include its relatively narrow radiation pattern, and the fact that it must be installed in a fixed, non-movable position.

4BI-3B.4 What is the characteristic impedance at the input of a nonresonant rhombic antenna?

A. 50 to 55 ohms   B. 70 to 75 ohms

C. 300 to 350 ohms   D. 700 to 800 ohms

The answer is D. The ideal transmission line to feed a rhombic antenna is a 600 ohm line.

4BI-3C.1 What is the effect of a <u>terminating resistor</u> on a rhombic antenna?

A. It reflects the standing waves on the antenna elements back to the transmitter

B. It changes the radiation pattern from essentially bidirectional to essentially unidirectional

C. It changes the radiation pattern from horizontal to vertical polarization

D. It decreases the ground loss

The answer is B. Without the terminating resistor, the rhombic antenna is resonant and has a bidirectional radiation pattern. When we terminate the antenna with a resistor, the antenna becomes non-resonant and the radiation pattern is unidirectional.

4BI-3C.2 What should be the value of the <u>terminating resistor</u> on a rhombic antenna?

A. About 50 ohms   B. About 75 ohms

C. About 800 ohms   D. About 1800 ohms

The answer is C. The input impedance of the rhombic antenna is about 600 to 800 ohms. When we terminate the antenna at the opposite end, it should be with a slightly larger non-inductive resistance. 800 ohms is a good choice. If a large non-inductive resistance is not available, several smaller ones will do. However, the resistors should be hooked up in such a way that their resultant resistance is 800 ohms. The terminating resistance should be able to handle about one-half of the power output of the transmitter.

4BI-4A.1 What factors determine the receiving antenna gain required at an amateur station in earth operation?

A. Height, transmitter power and antennas of satellite

B. Length of transmission line and impedance match between receiver and transmission line

C. Preamplifier location on transmission line and presence or absence of RF amplifier stages

D. Height of earth antenna and satellite orbit

The answer is A. Earth operation is earth-to-space-to-earth amateur radiocommunication by means of radio signals automatically retransmitted by stations in space operation.

The question refers to the receiving antenna gain REQUIRED at the ground station, and this is naturally affected by the height, transmitter power and antennas of the specific satellite that the ground station is working with. If the satellite is transmitting with very high power and the antennas have very high gain, then the receiver's antenna gain need not be as high as it would have to be if the satellite power and antenna gain were low.

4BI-4A.2 What factors determine the EIRP required by an amateur station in earth operation?
A. Satellite antennas and height, satellite receiver sensitivity
B. Path loss, earth antenna gain, signal–to–noise ratio
C. Satellite transmitter power and orientation of ground receiving antenna
D. Elevation of satellite above horizon, signal–to–noise ratio, satellite transmitter power

The answer is A. EIRP stands for EFFECTIVE ISOTROPIC RADIATED POWER. It is equal to the power being fed to the antenna, multiplied by the gain (with respect to an isotropic antenna) of the antenna. It should be obvious that the satellite antenna gain and receiver gain will affect the amount of Effective Radiated Power that is required of the ground station.

4BI-4A.3 What factors determine the EIRP required by an amateur station in telecommand operation?
A. Path loss, earth antenna gain, signal–to–noise ratio
B. Satellite antennas and height, satellite receiver sensitivity
C. Satellite transmitter power and orientation of ground receiving antenna
D. Elevation of satellite above horizon, signal–to–noise ratio, satellite transmitter power

The answer is B. Telecommand operation is earth–to–space amateur radio communication to initiate, modify, or terminate functions of a station in space operation. The same factors are involved in telecommand operation as in earth operation. See 4BI-4A.1 and 4BI-4A.2.

4BI-4A.4 How does the gain of a parabolic dish type antenna change when the operating frequency is doubled?
A. Gain does not change   B. Gain is multiplied by 0.707
C. Gain increases 6 dB   D. Gain increases 3 dB

The answer is C. The formula for the gain of a parabolic dish type antenna is:

$$G = k \left(\frac{\pi D}{\lambda} \right)^2$$

where: G is the power gain over an isotropic antenna, k is
an efficiency factor, D is the dish diameter in feet,
and λ is the wavelength in feet.

Doubling the frequency is the same as cutting the wavelength in half. From the formula, it can be seen that as the wavelength is cut in half, the power gain is quadrupled. This comes about because the wavelength is in the denominator of the equation and it is squared. Quadrupling the gain in terms of power is the same as a 6 dB increase. This can be seen from the basic dB power formula.

$$\text{Power gain} = 10 \log \frac{P_2}{P_1} = 10 \log 4 = 10 \times .6 = 6 \text{ dB}$$

See Figure 4BI-4A.4 for a diagram of a parabolic dish antenna.

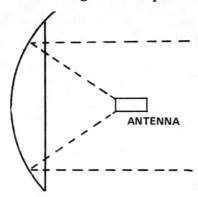

ANTENNA

Fig. 4BI-4A.4. A parabolic dish type antenna.

4BI-4B.1 What happens to the beamwidth of an antenna as the gain is increased?
A. The beamwidth increases geometrically as the gain is increased
B. The beamwidth increases arithmetically as the gain is increased
C. The beamwidth is essentially unaffected by the gain of the antenna
D. The beamwidth decreases as the gain is increased

The answer is D. See the discussion and figure in 4BI-4B.2. Increased antenna gain means that the radiated energy is being concentrated in a narrower beam, thus decreasing the beamwidth.

4BI-4B.2 What is the beamwidth of a symmetrical pattern antenna with a gain of 20 dB as compared to an isotropic radiator?
A. 10.1 degrees   B. 20.3 degrees   C. 45.0 degrees   D. 60.9 degrees

The answer is B. Antenna beamwidth is a measure of the directivity or sharpness of an antenna. It is the number of degrees between two points on the main lobe where the radiated power has decreased to one-half of the maximum power. See Figure 4BI-4B.2.

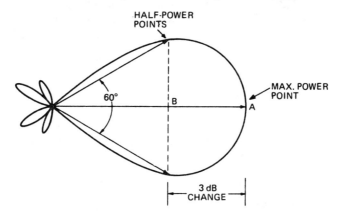

HALF-POWER
POINTS

60°

B

A

MAX. POWER
POINT

3 dB
CHANGE

Fig. 4BI-4B.2. Beamwidth measured at half power points.

In order to find the beamwidth of an antenna that produces a symmetrical pattern, we use the following formula:

$$\text{Beamwidth} = \frac{203}{\sqrt{10^{G/10}}}$$

where G is the gain of the antenna in dB

We then substitute the gain in the formula and solve the equation.

$$\text{Beamwidth} = \frac{203}{\sqrt{10^{20/10}}} = \frac{203}{\sqrt{10^2}} = \frac{203}{10} = 20.3 \text{ degrees}$$

4BI-4B.3 What is the beamwidth of a symmetrical pattern antenna with a gain of 30 dB as compared to an isotropic radiator?
A. 3.2 degrees   B. 6.4 degrees   C. 37 degrees   D. 60.4 degrees
   The answer is B. We use the same method to solve this problem as we used in answer 4BI-4B.2.

$$\text{Beamwidth} = \frac{203}{\sqrt{10^{30/10}}} = \frac{203}{\sqrt{10^3}} = \frac{203}{31.62} = 6.4 \text{ degrees}$$

4BI-4B.4 What is the beamwidth of a symmetrical pattern antenna with a gain of 15 dB as compared to an isotropic radiator?
A. 72 degrees   B. 52 degrees   C. 36.1 degrees   D. 3.61 degrees
   The answer is C. We use the same method to solve this problem as we used in answer 4BI-4B.2.

$$\text{Beamwidth} = \frac{203}{\sqrt{10^{15/10}}} = \frac{203}{\sqrt{10^{1.5}}} = \frac{203}{\sqrt{31.62}} = 36.1 \text{ degrees}$$

4BI-4B.5 What is the beamwidth of a symmetrical pattern antenna with a gain of 12 dB as compared to an isotropic radiator?
A. 34.8 degrees   B. 45.0 degrees   C. 58.0 degrees   D. 51.0 degrees
   The answer is D. We use the same method to solve this problem as we used in answer 4BI-4B.2.

$$\text{Beamwidth} = \frac{203}{\sqrt{10^{12/10}}} = \frac{203}{\sqrt{10^{1.2}}} = \frac{203}{\sqrt{15.85}} = 51 \text{ degrees}$$

4BI-4C.1 How is circular polarization produced using linearly-polarized antennas?
A. Stack two Yagis, fed 90 degrees out of phase, to form an array with the respective elements in parallel planes
B. Stack two Yagis, fed in phase, to form an array with the respective elements in parallel planes
C. Arrange two Yagis perpendicular to each other, with the driven elements in the same plane, and fed 90 degrees out of phase

D. Arrange two Yagis perpendicular to each other, with the driven elements in the same plane, and fed in phase

The answer is C. Amateur satellites tumble while in orbit. This causes the polarization of signals from the satellite to change. In addition, the polarization is further altered as the signal travels through the earth's atmosphere (this phenomenon, which increases with increasing frequency, is known as Faraday rotation). For the amateur using either a vertically or horizontally polarized antenna, the effect is that signals are received at a strong level, only to disappear into the noise a few seconds later as the polarization shifts. In order to overcome the effects of polarization, a circularly polarized antenna should be used. When we say that a signal has circular polarization, we mean that it has both vertical and horizontal components, and can be picked up by both vertically and horizontally polarized antennas.

There are many ways of generating a circular polarized wave. In one method, two dipole or Yagi antennas are mounted at right angles to each other and fed RF energy 90 degrees out of phase.

4BI-4C.2 Why does an antenna system for earth operation (for communications through a satellite) need to have rotators for both azimuth and elevation control?
A. In order to point the antenna above the horizon to avoid terrestrial interference
B. Satellite antennas require two rotators because they are so large and heavy
C. In order to track the satellite as it orbits the earth
D. The elevation rotator points the antenna at the satellite and the azimuth rotator changes the antenna polarization

The answer is C. Rotators are necessary since the satellite is continuously moving.

4BI-5.1 What term describes a method used to match a high- impedance transmission line to a lower impedance antenna by connecting the line to the driven element in two places, spaced a fraction of a wavelength on each side of the driven element center?
A. The gamma matching system   B. The delta matching system
C. The omega matching system   D. The stub matching system

The answer is B. The DELTA MATCH is a system in which the ends of the transmission line are fanned out and then connected to

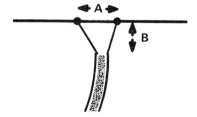

Figure 4BI-5.1. A delta match.

the dipole (see Figure 4BI-5.1). It is used to connect a transmission line of any impedance to a half-wave antenna. The fanning out of the transmission line changes its impedance so that it can match the points on the antenna that it is connected to. The delta match can be used with a parallel conductor transmission line, and it can also be used with a coax line if a balun is added. In order to obtain a proper impedance match and a low SWR, the correct dimensions of A and B must be closely adhered to.

4BI-5.2 What term describes an unbalanced feed system in which the driven element is fed both at the center of that element and a fraction of a wavelength to one side of center?
A. The gamma matching system   B. The delta matching system
C. The omega matching system   D. The stub matching system

The answer is A. The GAMMA MATCH is used primarily to match an unbalanced coaxial transmission line to a dipole antenna. It contains one variable capacitor, one adjustable clamp and a small length of tubing. See Figure 4BI-5.2.

The length of the tubing is usually fixed, while the sliding clamp is used to adjust dimension A for minimum SWR. The capacitor is used to tune out the reactance of the matching section.

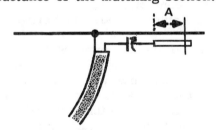

Figure 4BI-5.2. A gamma match.

4BI-5.3 What term describes a method of antenna impedance matching that uses a short section of transmission line connected to the antenna feed line near the antenna and perpendicular to the feed line?
A. The gamma matching system   B. The delta matching system
C. The omega matching system   D. The stub matching system

The answer is D. The "stub system" is a means of impedance matching between an antenna and a transmission line. It also reduces the standing wave ratio. A "stub" is a section of transmission line, typically 1/4 or 1/2 wavelength. There are many different systems used in stub tuning. Sometimes the stub is connected to the transmission line at a point near the antenna. At other times, the stub is connected directly to the antenna and the transmission line is connected to a point on the stub.

Figure 4BI-5.3 shows a 1/2 wave dipole antenna connected to a transmission line via a 1/4 wave matching stub. In this case, the stub is unterminated. At other times, the stub is shorted. The transmission line is moved between the top and bottom of the stub until a point is chosen, which is exactly the same as the characteristic impedance of the transmission line. Thus, the transmission line will be terminated at an impedance equal to its own characteristic impedance, and the stub will feed the energy to the antenna proper. There is no mismatch between the stub and the 1/2 wave part of the antenna because the stub is actually a part of the total antenna. Since the line is properly terminated, there will be no reflected waves. Hence, there will be no standing waves and all of the energy will be radiated.

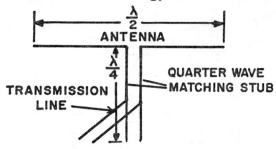

Figure 4BI-5.3. Antenna matching with a stub.

4BI-5.4 What should be the approximate capacitance of the resonating capacitor in a gamma matching circuit on a 1/2- wavelength dipole antenna for the 20-meter band?

A. 70 pF   B. 140 pF   C. 200 pF   D. 0.2 pF

The answer is B. The value of the resonating capacitor should be about 7 pf per meter of wavelength. Therefore, a 20 meter dipole should have a gamma matching capacitor of about 20 x 7 = 140 pf. The capacitor is variable and is adjusted in the tune-up procedure for minimum SWR.

4BI-5.5 What should be the approximate capacitance of the resonating capacitor in a gamma matching circuit on a 1/2- wavelength dipole antenna for the 10-meter band?

A. 70 pF   B. 140 pF   C. 200 pF   D. 0.2 pF

The answer is A. See answer 4BI-5.4. A 10 meter dipole should have a gamma matching capacitor of 10 X 7 = 70 pf.

4BI-6A.1 What kind of impedance does a 1/8-wavelength transmission line present to a generator when the line is shorted at the far end?

A. A capacitive reactance
B. The same as the characteristic impedance of the line
C. An inductive reactance
D. The same as the input impedance to the final generator stage

The answer is C. A resonant transmission line, like a tuned circuit, is resonant at some particular frequency. A resonant transmission line will present, to its source of energy, a high or low resistive impedance at multiples of a quarter wavelength. Whether the impedance is high or low at these points, depends on whether the line is short or open-circuited at the output end. At points that are not exact multiples of a quarter wavelength, the line acts as a capacitor or an inductor.

A 1/8 wavelength is not an exact multiple of a quarter of a wavelength. It therefore acts as a capacitor or inductor. If the far end of the line is shorted, the generator sees an inductive reactance whose value is equal to the transmission line impedance. Table 4BI-6A.1 gives the input impedances for the various transmission line lengths and terminations. Note that for a 1/8 line shorted at the far end, the generator sees an inductive reactance which is numerically equal to the characteristic impedance of the line.

Transmission line length in wavelengths	Termination	INPUT IMPEDANCE Amount	Type
one eighth	open	equal to Z_0	capacitive reactance
one eighth	shorted	equal to Z_0	inductive reactance
one quarter	open	very low	resistive
one quarter	shorted	very high	resistive
three eighths	open	equal to Z_0	inductive reactance
three eighths	shorted	equal to Z_0	capacitive reactance
one half	open	very high	resistive
one half	shorted	very low	resistive
any length	Z_0	equal to Z_0	resistive

Z_0 indicates characteristic line impedance.

Fig. 4BI-6A.1. Resonant line input impedances.

4BI-6A.2 What kind of impedance does a 1/8-wavelength transmission line present to a generator when the line is open at the far end?
A. The same as the characteristic impedance of the line
B. An inductive reactance
C. A capacitive reactance
D. The same as the input impedance of the final generator stage
    The answer is C. See answer and Table 4BI-6A.1. Note that for a 1/8 line open at the far end, the generator sees a capacitive reactance numerically equal to the characteristic impedance of the line.

4BI-6B.1 What kind of impedance does a 1/4-wavelength transmission line present to a generator when the line is shorted at the far end?
A. A very high impedance
B. A very low impedance
C. The same as the characteristic impedance of the transmission line
D. The same as the generator output impedance
    The answer is A. A quarter-wave section of a transmission line exhibits different properties, depending upon its termination. Figure 4BI-6B.1A illustrates a quarter-wave section of a line terminated in a pure resistance that is equal to the characteristic impedance of the line. RF energy travels along the line and is absorbed by the load. There are no reflections or standing waves. It is as though the line were infinitely long. An ammeter or voltmeter, inserted anywhere along the line, would show the same current and voltage. The impedance is also the same at any point in the line because the ratio of E to I

is the same.

In Figure 4BI-6B.1B, the quarter-wave line is terminated by a short circuit. The RF energy travels along the line till it meets the short, and is then reflected back to the input. Standing waves will appear. At the short circuit, the current is high, the voltage is low and the impedance is low. The opposite is true at the input end. The current is low, the voltage is high and the impedance is infinite. Thus, the input appears as an infinite impedance to a generator. See Table 4BI-6A.1.

Figure 4BI-6B.1C shows a line whose termination is open. The RF energy travels to the end and is reflected back to form standing waves. Since the end is open, the current is zero and the voltage and impedance are high. At the input, the current is high and the voltage and impedance are low. Thus, the input appears as a very low impedance to a generator.

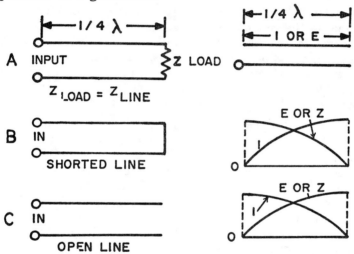

Fig. 4BI-6B.1. Quarter-wave sections of a transmission line

4BI-6B.2 What kind of impedance does a 1/4-wavelength transmission line present to a generator when the line is open at the far end?
A. A very high impedance   B. A very low impedance
C. The same as the characteristic impedance of the line
D. The same as the input impedance to the final generator stage
The answer is B. See answer 4BI-6B.1 and table 4BI-6A.1.

4BI-6C.1 What kind of impedance does a 3/8-wavelength transmission line present to a generator when the line is shorted at the far end?
A. The same as the characteristic impedance of the line
B. An inductive reactance
C. A capacitive reactance
D. The same as the input impedance to the final generator stage
The answer is C. See answer and Table 4BI-6A.1. A 3/8 wavelength transmission line, terminated with a short circuit, behaves like a capacitor whose reactance is equal to the characteristic impedance of the transmission line.

4BI-6C.2 What kind of impedance does a 3/8-wavelength transmission line present to a generator when the line is open at the far end?
A. A capacitive reactance
B. The same as the characteristic impedance of the line
C. An inductive reactance
D. The same as the input impedance to the final generator stage
    The answer is C. See answer and Table 4BI-6A.1. A 3/8 wavelength transmission line, terminated in an open circuit, behaves like an inductor whose reactance is equal to the characteristic impedance of the transmission line.

4BI-6D.1 What kind of impedance does a 1/2-wavelength transmission line present to a generator when the line is shorted at the far end?
A. A very high impedance   B. A very low impedance
C. The same as the characteristic impedance of the line
D. The same as the output impedance of the generator
    The answer is B. See answer and Table 4BI-6A.1. A one-half wavelength transmission line, terminated with a short circuit, behaves like a series resonant circuit. The current is maximum, the impedance is minimum, and since the reactances cancel each other out, the only thing that remains is the low value of resistance.

4BI-6D.2 What kind of impedance does a 1/2-wavelength transmission line present to a generator when the line is open at the far end?
A. A very high impedance   B. A very low impedance
C. The same as the characteristic impedance of the line
D. The same as the output impedance of the generator
    The answer is A. See answer and Table 4BI-6A.1. A one-half wavelength transmission line, terminated in an open circuit, behaves like a parallel resonant circuit. It has a very high impedance, and the current flow through the circuit is minimum.

FREQUENCY ALLOCATIONS FOR POPULAR AMATEUR BANDS
All in MegaHertz. "X" indicates no privileges.

CLASSES	NOVICE		TECHNICIAN		GENERAL AND CONDITIONAL		ADVANCED		EXTRA	
BANDS	CW	PHONE	CW	PHONE	CW	PHONE	CW	PHONE	CW	PHONE
80 METERS	3.7 to 3.75	X	3.7 to 3.75	X	3.525 to 3.750 and 3.85 to 4.0	3.85 to 4.0	3.525 to 3.750 and 3.775 to 4.0	3.775 to 4.0	3.5 to 4.0	3.75 to 4.0
40 METERS	7.1 to 7.15	X	7.1 to 7.15	X	7.025 to 7.150 and 7.225 to 7.3	7.225 to 7.3	7.025 to 7.3	7.15 to 7.3	7.0 to 7.3	7.15 to 7.3
20 METERS	X	X	X	X	14.025 to 14.15 and 14.225 to 14.35	14.225 to 14.35	14.025 to 14.15 and 14.175 to 14.35	14.175 to 14.35	14.0 to 14.35	14.15 to 14.35
15 METERS	21.1 to 21.2	X	21.1 to 21.2	X	21.025 to 21.20 and 21.30 to 21.450	21.3 to 21.450	21.025 to 21.20 and 21.225 to 21.45	21.225 to 21.450	21.0 to 21.450	21.2 to 21.45
10 METERS	28.1 to 28.5	28.3 to 28.5	28.1 to 28.5	28.3 to 28.5	28.0 to 29.7	28.3 to 29.7	28.0 to 29.7	28.3 to 29.7	28.0 to 29.7	28.3 to 29.7
6 METERS	X	X	50.0 to 54.0	50.1 to 54.0	50.0 to 54.0	50.1 to 54.0	50.0 to 54.0	50.1 to 54.0	50.0 to 54.0	50.1 to 54.0
2 METERS	X	X	144.0 to 148.0	144.1 to 148.0	144.0 to 148.0	144.1 to 148.0	144.0 to 148.0	144.1 to 148.0	144.0 to 148.0	144.1 to 148.0

Table of Emissions

The FCC is now using the new WARC emission symbols. In the new system, the old 2-character symbols have been replaced with 3-character symbols. The 3-character symbols give more specific information concerning the emissions that they represent.

FIRST CHARACTER
N   Emission of an unmodulated carrier
A   AM double-sideband
J   Single sideband, suppressed carrier
F   Frequency modulation
P   Sequence of unmodulated pulses
C   Vestigial sidebands

SECOND CHARACTER
0   No modulating symbol
1   Digital information - no modulation
2   Digital information with modulation
3   Modulated with analog information

THIRD CHARACTER
N   No information transmitted
A   Telegraphy for reception by air
B   Telegraphy for automatic reception
C   Facsimile
D   Data transmission, telemetry, telecommand
E   Telephony
F   Television

Traditional Symbol		New Symbol
AMPLITUDE MODULATED		
Unmodulated	A0	N0N
Keyed on/off	A1	A1A
Tones keyed on/off	A2	A2A
AM data		A2D
Keyed tones w/SSB	A2J	J2A
SSB data		J2D
AM voice	A3	A3E
Voice w/SSB	A3J	J3E
AM facsimile	A4	A3C
SSB television	A5	C3F
AM television	A5	A3F
FREQUENCY MODULATED		
Unmodulated	F0	N0N
Switched between		
two frequencies	F1	F1B
Switched tones	F2	F2A
FM data		F2D
FM voice	F3	F3E
FM facsimile	F4	F3C
FM television	F5	F3F
PULSE MODULATED		
Phase	P	P1B

THE COMPLETE MORSE CODE
COURSE FOR THE PC

*Generate random characters at ANY speed

*Generate random QSO's-similar to the VEC

  exams - at ANY speed

*Send text from any external data file

*Complete lesson on learning the Morse Code

*Includes 32 page book on Code learning and user's

  manual

*Plus many, many more features

Ameco's Morse Code course for the PC is the most versatile program of its kind. It is user friendly and menu driven, containing over 18 options. It will run on any IBM PC/XT/AT (or 100   percent compatible) at any clock speed, in either monochrome or color.

There are many other features, including quiz sessions for the beginner, as well as the ability to alter letter, character and word spacing to simulate HI/LO spacing. The program also will turn your keyboard into a straight or iambic keyer.

This course is ideal for the beginner, and perfect for the licensed ham who wishes to upgrade! All from AMECO Publishing, the oldest and largest publisher of code training material, for over 38 years.

AMECO'S Morse Code Course for the PC (Cat. No. 107-PC)..$19.95

STOP TV, FM & VCR INTERFERENCE

with the new AMECO

HIGH PASS FILTERS

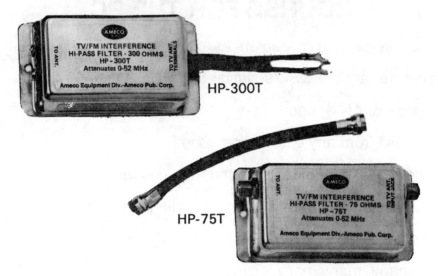

HP-300T

HP-75T

Eliminate interference to TV, FM and VCR sets caused by Amateur Radio, CB, shortwave, police, medical equipment etc. with high quality, lab-type filters. Each filter has 9 shielded sections, a sharp cut-off at 52 MHz. and over 70 dB attenuation below 50 MHz.

Model HP-75T is for 75 ohm coax installations. It is also perfect for the new cable systems. Installation is simple because all connectors and a length of coaxial cable are included. $12.95

Model HP-300T is for 300 ohm twin-lead systems. Installation is simple because the filter input has screw terminals to receive the antenna twin-lead and the filter output has twin-lead with spade lugs to connect to the TV set. $12.95

AMECO EQUIPMENT CO.
div. AMECO PUBLISHING CORP.
220 East Jericho Tpke.
Mineola, New York 11501
Tel # (516)-741-5030

TUNABLE PREAMPLIFIER ANTENNA
Model TPA

- Can be used as a tunable preamplifier or an indoor active antenna.

- Complete coverage from 0.22 to 30 MHz.

- Improves receiver gain and noise figure.

- Over 20 dB gain on all frequencies.

Model TPA is a dual function unit. It can be used as a preamplifier to improve the gain of a receiver, or as an indoor active antenna when an outdoor antenna is not available.

Model TPA contains a tuned RF amplifier that covers all frequencies from 0.22 to 30 MHz., including amateur bands, all foreign broadcast bands, citizen's band and all other services within this range. A dual gate, field effect transistor provides an excellent noise figure and over 20 dB gain. The weak signal performance of most receivers is improved.

Model TPA uses either an internal 9 volt battery or an AC adapter, such as Ameco Model P-9T. As a preamplifier, the input matches most antennas. Long wire, 300 ohm and random length antennas can also be used with good results.

When no external antenna is available, the preamplifier's whip antenna gives excellent results.

Model TPA. Preamplifier/Antenna $76.95
Model P-9T. AC adapter for TPA .... $ 8.50

Ameco Equipment Div. of Ameco Publishing Corp.
220 East Jericho Turnpike, Mineola, NY 11501
(516) 741-5030

NEW PREAMPLIFIER FOR TRANSCEIVER USE - MODEL PT-3

Model PT-3 is a continuously tunable 6-160 meter pre-amplifier, specifically designed for use with a transceiver. It features a dual-gate FET transistor amplifier which provides a low noise figure, thus improving the sensitivity of the receiver section of the transceiver. Signals are increased as much as 26 dB. A unique built-in RF sensing circuit enables the PT-3 to bypass itself when the transceiver is transmitting. The gain of the preamplifier can be varied by the front panel RF gain control. This control can also be used to reduce the amplification prior to the first mixer, thereby minimizing or eliminating overload effects caused by strong off-channel signals. The PT-3 has an adjustable delay control on the front panel that determines the amount of time the PT-3 will stay in the transmit mode before returning to the receive mode after the operator stops talking.

Provisions are included so that the PT-3 can easily be modified to feed the input of a second receiver. The PT-3 can also be modified by the user so that a separate receiving antenna can be used in addition to the main antenna.

The input and output impedances of the PT-3 are nominally 50 ohms. This matches most types of antennas.

12 volts DC is required to power the PT-3. If 120 volts AC is available, the Ameco Adapter, Model P-12T, can be used. It plugs into the 120 volt AC source and delivers an output of 12 volts DC for the standard PT-3.

Cat. #PT-3 Preamplifier for 160-6 meters. Wired & Tested.

Cat. #P-12T Adapter-120 V.AC to 12V.DC. Wired & Tested.

- **Improves sensitivity while receiving**

- **No modification required to transceiver**

- **For single sideband, AM or CW use**

- **Can handle transceiver output up to 350W.**

- **Can be used with separate linear amplifiers**

- **For transceiver or receiver use**

- **Adjustable delay control on front panel**

- **Optional second receiver capability**

- **Optional separate receiving antenna capability**

Cat. #PT-3......$114.95   Cat. #P-12T.....$8.95

Ameco Equipment Div. of Ameco Publishing Corp.
220 East Jericho Turnpike, Mineola, NY 11501
(516) 741-5030